The Legendary
MUSCLE CAR

The Legendary
MUSCLE
CAR

JIM GLASTONBURY

CHARTWELL
BOOKS

Quarto is the authority on a wide range of topics.

Quarto educates, entertains and enriches the lives of our readers—enthusiasts and lovers of hands-on living.

www.quartoknows.com

This edition published in 2016 by
CHARTWELL BOOKS
an imprint of Book Sales
a division of Quarto Publishing Group USA Inc.
142 West 36th Street, 4th Floor
New York, NY 10018
USA

ISBN-13: 978-0-7858-3479-3

10 9 8 7 6 5 4 3 2 1

Printed in China

For all editorial enquiries, please contact:
www.regencyhousepublishing.com

CONTENTS

INTRODUCTION

What is a muscle car? First of all, let us eliminate what it is not: it is not a piece of Italian exotica, a Ferrari or a Lamborghini, cars which are just too complex and too specialized; nor is it a German Porsche, which is too efficient and too clever by half; nor yet a classic British sports car, a Morgan, TVR, or Jaguar, which could never be regarded as fitting the bill. Sports cars, by and large, are not muscle cars, with two notable exceptions: the legendary AC Cobra of the 1960s, and the Dodge Viper of the 1990s. These followed the muscle car creed of back-to-basics raw power.

In effect, muscle cars always were, and always will be, a

LEFT: The AC Cobra, the English muscle car.

OPPOSITE: Think GTO, think muscle car!

OVERLEAF: A Mustang Drag Racer.

*quintessentially North American phenomenon. The basic concept is
something like this: take a mid-sized American sedan, nothing complex,
upmarket, or fancy, in fact the sort of car one would use to collect the
groceries in any American town on any day of the week; add the biggest,
raunchiest V8 that it is possible to squeeze under the hood; and there it is.*

*The muscle car concept really is as simple as that. Moreover, the young
men who desired these cars, and most of them were young and men, though
that would change, were not interested in technical sophistication, nor
handling finesse, nor even top speed. Cubic inches, horsepower and
acceleration rates were the only figures that counted. Muscle cars were loud,
proud, and in your face, and did not pretend to be anything else. They might
have been simple, even crude, but for roaring, pumping, tire-smoking
standing starts, they were the business. To an American youth culture raised
on drag racing, red-light street racing, and hot-rodding, they were irresistible.*

*The "Big Three" manufacturers soon woke to this fact, and joined the
power race to offer more cubic inches, more horsepower, and fewer
seconds over the standing quarter. For a few short years, between 1965
and 1970, it seemed as though the race would never end. The result was
often more power than the car (and the driver) could handle safely, but
then part of the attraction was making a four-seater sedan go faster than it
was ever intended.*

*But the situation could not last. The combination of high horsepower in
the hands of young drivers saw accident rates soar, and insurance premiums
followed suit. Moreover, the climate of the times was changing, with a whole*

raft of safety and emissions legislation coming into force in the late 1960s and early 1970s. So even before the first oil crisis made itself felt, the first-generation muscle cars were already on their way out. By the 1980s, however, they were beginning to creep back, first with turbocharged fours, then V8s; by the 1990s, muscle cars were back with a vengeance: more "high-tech" than before, even sophisticated, with ABS, electronic fuel injection, and multi-valve engines. Manufacturers were by now talking virtuously about catalytic converters and air bags, but the truth was that performance was selling once again. Anti-social? Yes. Irresponsible? Of course. But one thing was certain, the muscle car was back.

PONTIAC GTO

The Pontiac GTO was the world's first true muscle car, which is the consensus opinion of just about everyone who has written on the subject. Of course, there were rumbling, grumbling, growling V8 sedans well before the GTO came along, but most were big, heavy, full-sized cars. The difference was that the GTO married one of those full-sized V8s with a lighter, intermediate bodyshell. The result was stunning performance in an affordable package, and the muscle car was born. Pontiac could actually have sold more than it built in 1964, and it was soon apparent that it was a new class of car. The odd thing was that it nearly didn't happen at all.

There was no long-term plan behind the GTO, no systematic market research or seven-year research and development program. The GTO came about because a few hot car fanatics were in the right place at the right time, when it could so easily

OPPOSITE: A 1969 Pontiac GTO.

ABOVE: The original muscle car which started it all, the Pontiac GTO.

OVERLEAF: A 1968 GTO/Tempest.

have been the wrong time. In 1963, General Motors decreed that none of its divisions would go racing any more, or even produce overtly sporting cars. Pontiac was also restricted to a 300-cu in (4.92-liter) capacity ceiling on standard engines in intermediate cars. It was a bitter blow, as of all the General Motors marques Pontiac, like Chevrolet, prided itself on building performance cars. But in the GM pecking order, it did not have the resources of Chevrolet to develop an all-new car to get around this decision.

It was even worse news for key Pontiac men like chief engineer John DeLorean, general manager Elliot "Pete" Estes and his predecessor Semon "Bunkie" Knudsen, engineers Russ Gee and Bill Collins, and advertising man Jim Wangers. All were hot car enthusiasts, convinced that the way to sell cars was by making image-building hot-rods. Fortunately for their fellow enthusiasts, all were in a position to do something about it.

Of course, Pontiac already made fast full-sized saloons, and Knudsen, Estes, and DeLorean in particular had transformed the division's image from staid to sporty in a few years. Super Duty engine tuning parts were aimed at NASCAR racing and the drag strip, but super-stockers also rolled off the production line, road-legal and ready to go.

They were certainly fast, but also loud, bad-tempered, and decidedly spartan in comfort, while high-lift, long-duration solid-lifter cams and multiple carburettors made them temperamental. In an effort to reduce weight, aluminium or fiberglass body panels were often fitted: Pontiac even drilled out the chassis of the Super Duty Catalina to trim further "fat" from the vehicle. These were specialized racing machines that took time to assemble, so they were not cheap. There was one other thing: a prospective buyer could not just stroll into his local Pontiac dealer and order one for the grocery run. In the words of author Thomas DeMauro, one had to have the right credentials: "Most required a National Hot Rod Association (NHRA) license, a letter of recommendation from God and compromising photos of the auto maker's general manager driving a Volkswagen to get one."

This was where the GTO broke new ground. It used a standard, smooth, reliable, and docile 389-cu in (6.375-liter) V8 straight out of Pontiac's parts inventory. But it was fast (and this was the master stroke) because it was shoehorned into an intermediate saloon, the Tempest, that weighed a whole 400lb (181kg) less than a full-sized car. So it could do without rattly fiberglass panels, did not need the care and

"Muscle cars are loud, proud, and in your face, with no other pretensions than to be just that."

For 1968, the GTO Tempest acquired vertically-stacked headlamps.

attention of a dedicated mechanic, and could afford to include all the hot car accessories like bucket seats and wide tires without upsetting the weight balance. Best of all, being a mix-and-match of existing parts that slotted neatly together on the production line, it was cheap. The GTO started at $2,700, which made for truly affordable performance.

This was the muscle car concept (a big V8 in a smaller saloon) and the GTO really did start it all. As it happened, the concept was easier for Pontiac to put into practice than anyone else, though was more by accident than design. Fitting a big-block V8 into a smaller car was not as easy as it sounds, as the big-blocks from Ford and Chrysler, not to mention Pontiac's in-house rivals, were physically bigger than the small-blocks, so one could not merely take one out and slot the other in. All Pontiac V8s, on the other hand, from the 326 up to the big 421 were physically the same size, so it was relatively simple to swap them around. That made a car like the GTO a relatively quick and easy variation to develop.

Better still, all those Pontiac V8s lost some weight in 1963. Without careful development, slotting a weighty big-block V8 into an intermediate car could result in a nose-heavy monster that would understeer at the slightest provocation. Pontiac's 389 was now light enough to minimize this problem. There was something else too. The unit-construction Pontiac Tempest of 1963 was not up to handling the power and torque of the 389. But for 1964, it was redesigned with a conventional perimeter frame, four-link rear suspension and solid rear axle, which could easily take a beefier output. Same-size engines, less weight, tougher chassis: it was almost as though the GTO was really meant to be.

Threading the Loops

But there was still one small problem: officially, no middleweight Pontiac could be offered with an engine larger than 300 cu in unless, of course, it was an option. So when it was announced in October 1963, the GTO was not a model in its own right, but a $295 option package which could be ordered with the LeMans Tempest in coupé, hardtop, or convertible forms. The clever thing was that the GTO option was not just a big engine, but a whole range of detail parts that made the GTO a model in its own right. It was sneaky, but it worked, getting around the 300-cu in ceiling and kicking off a legend all at one and the same time.

The name, incidentally, stood for Gran Turismo Omologato, which was another piece of marketing sleight of hand. It implied that Pontiac had applied to the FIA (Fédération Internationale de l'Automobile, the world governing body for motor sports) to have the car homologated for racing. That is why the legendary Ferrari GTO was so named. Some people were horrified that this parts-bin special should take on such a hallowed badge without earning the right. Others could not have cared less. Either way, the controversy simply created more publicity for Pontiac in general and the GTO in particular: like a celebrity publishing his warts and all autobiography, Pontiac could not lose.

So what did one get for the $295? The basic engine was Pontiac's existing 389-cu in (6.375-liter) V8, which in standard tune produced 325hp (242kW) at 4,800rpm. The engine was not totally "off the shelf," however: the standard heads were replaced with 421 HO items, with big valves, and allowing a 10.75:1 compression with the flat-top pistons, and other changes included heavy-duty valve springs and a Carter four-barrel carburettor, plus different lifters and camshaft. If 325hp was not enough, one could change the option and specify Tri-Power, three Rochester two-barrel carburettors in place of the single four-barrel, which produced 348hp (259.5kW) at 4,900rpm. The triple carburettor set-up was a hangover from Pontiac's Super Duty days. Either way, one achieved eye-popping performance.

In January 1964 *Motor Trend* tested a GTO, a four-speed convertible which rocketed to 60mph in 7.7 seconds and covered the standing quarter-mile in 15.8 seconds. Meanwhile, *Car and Driver* published a now famous test between the GTO

and a genuine Ferrari GTO. It was a nice idea, except that in its eagerness to out-GTO the Ferrari, Pontiac supplied two test cars, both fitted with tuned 421-cu in (6.9-liter) engines that managed 0–100 mph (161km/h) in 11 seconds (which *C&D* headlined on its front cover) and turned in a standing quarter of 13.1 seconds. It really made a nonsense of the whole point of road-testing cars: no one could walk into his Pontiac dealer and order a GTO like that. On the other hand, it worked for

OPPOSITE, ABOVE, OVERLEAF & PAGES 22–23: The GTO was really a Tempest with some hot options. This is a 1966 model.

Pontiac. *Car and Driver* declared that in many ways the U.S.-built GTO was better than the Italian vehicle. This delighted and infuriated so many enthusiasts that the magazine was still getting mail on the subject two years later.

But in the real world, it was not just a case of choosing between two states of

tune: the GTO buyer could specify any of a bewildering array of other options, covering interior, exterior, and mechanics. It was fun, it helped to "personalize" the car, gave dealers and manufacturer extra profit mark-ups, and was always part of the muscle car buying experience. The basic transmission was a three-speed manual, with floor-mounted Muncie shift. Or one could have a wide-ratio or close-ratio four-speed, with at least half a dozen different rear axle ratios. There was an automatic too, a two-speed Super Turbine, which was upgraded for the GTO with a high-output governor to allow higher speed and rpm up-changes, as well as higher

clutch capacity, among other things. These detail changes to existing parts indicated that the GTO was not a hurried-through option, but well thought out and developed, notwithstanding what was said at the beginning of this chapter.

Suspension changes were limited to high-rate springs, specially valved shock absorbers, and longer rear stabilizers than standard, backed up by wider 6-in (15.2-cm) rim steel wheels and low-profile red-stripe tires. Not everything changed, though, and the 325-hp GTO carried the same 9.5-in (24.1-cm) drum brakes as the

six-cylinder Tempest; this would be a GTO weak spot for quite a while. The only concession to the GTO's 115-mph (185-km/h) top speed was the option of semi-metallic linings, said to reduce fade.

There was no chance of anyone mistaking the GTO for a pedestrian Tempest or LeMans, or for that matter an Oldsmobile Cutlass or F85, Buick Skylark, or Chevrolet Chevelle/Malibu, all of which used the same A-body platform. As well as those wider red-stripe tires, there were two fake hood scoops with chrome-ribbed trim and GTO badges on the grille, rear end, and flanks. Inside, bucket seats were standard, with an engine-turned trim on the dashboard. What one did not get was full instrumentation, and the standard set simply comprised a speedometer and fuel gauge, backed by various idiot lights. Keen drivers could specify extras, such as a semi-circular tachometer which indicated an optimistic maximum 7,000rpm. It also proved to be wildly inaccurate, and was prone to emit a steady hum, later fixed: it is now a collector's item.

So what did the buyers think of Pontiac's latest option pack, the GTO? They loved it. Despite being on sale for only part of the 1964 season, over 7,000 coupés, 6,600 convertibles, and over 18,000 hardtops were sold as GTOs, which was not bad for what was really just another option.

If there had been any annoyance among General Motors top brass at Pontiac's bare-faced cheek with the GTO, it did not show. Anything the senior management may have felt was mollified by the 75,000-plus GTOs that were sold in 1965. In any case, the gates were open now, and in its second year the GTO faced in-house GM competition from the Oldsmobile 4-4-2, Buick Gran Sport, and Chevrolet Malibu SS.

But one has a headstart, being first, and the GTO's first-year impact paid off in its second year. Pontiac made the most of that more aggressive front-end styling, with vertical twin headlamps and one large hood scoop in place of the two smaller ones.

Bolder front grilles and concealed taillights completed the exterior changes, while inside there was a new Rally gauge option which brought a proper full-sized tachometer as well as water temperature and oil pressure gauges. The brakes remained unchanged except for optional aluminium finned drums, more efficient at dissipating heat and so reducing fade. The optional wire wheel covers now had slots to allow cooling air in, but the real solution was to ditch the drums altogether in favour of power-assisted discs.

The big news was more power, a clear indication that the Pontiac performance restriction had been swept away by the GTO's success. Both four-barrel and Tri-Power V8s received better breathing from altered cylinder heads and inlet manifolds. That boosted the base engine to 335hp (250kW), enough for a 16.1-second quarter-mile, and 0–60mph in 7.2 seconds. The triple-carburettor Tri-Power was fitted with a new camshaft as well, with 288/302 degrees duration, to produce 360hp (268kW) at 5,200rpm. To promote its new star, Pontiac emphasized the big cat theme. "Have new tigers," went one, "need tamer. Apply at any Pontiac dealer." There was also the "GeeTO Tiger" record, revealing the sounds of a GTO being driven hard at the GM Proving Ground in Milford, Michigan. That was 50 cents, but if pocket money did not stretch quite that far, 25 cents bought a set of giant color photographs of GeeTO Tiger in action.

The year 1968 was a high point for the GTO, the year when it finally became a model in its own right, not simply an option package. Better still, sales were up 28 per cent on the previous year, with nearly 100,000 cars sold. As ever, the hardtop was by far the most popular, making up 75 per cent of those 100,000 vehicles. A substantial minority (just over 10,000) plumped for the coupé and nearly 13,000 GTO buyers opted for convertibles. But they were not all performance freaks: less than one in five paid extra for the 360-hp Tri-Tri-Power set-up.

Whatever the model mix, the GTO was undisputed king of the muscle cars in 1966, selling more than anyone else, if one ignores the pony cars. It was quite an achievement, as the marketplace was now crowded with a host of rivals trying to muscle in on the quick-car scene. Ford fitted a 390-cu in (6.394-liter) V8 to the Fairlane GT and GTA, while Mercury did the same, turning the Comet into the Cyclone. More seriously, both could be had with a 427-cu in (7.0-liter) option, and Chrysler's 426-cu in (6.98-liter) Hemi was just coming on stream. GM cars were at a decided disadvantage, as the parent decreed that no GM intermediate car could be sold with an engine bigger than 400 cu in (6.55 liters), though Buick got a 401-cu in (6.57-liter) dispensation. This time they would make it stick: there would be no sneaky big-engine options to get around it.

But this did not appear to harm GTO sales, which were aided no doubt by the

RIGHT, PAGES 28-29 & 30–33: The 1969/70 GTO had a front body-colored bumper, however the rear remained in chrome.

new Coke-bottle styling of that year. This little kick-up over the rear wheels, so ubiquitous in the late 1960s and early 1970s in cars built all over the world, rather than just in the U.S.A., had originated in Pontiac's full-sized cars of 1963. The hardtop got a semi-fastback, which combined rakish rear pillars with a more upright rear window. It looked good, but created a serious blind spot. Other changes were minor, though did include a Ram Air option for the Tri-Power engine. As well as the foam-sealed air pan, it included a longer duration camshaft (301/313 degrees) and heavy-duty valve springs. Strangely, it was quoted at the same 360hp as the standard Tri-Power. Either way, this was the final year for the complex three-carburettor set-up.

But did the GTO deserve its best-seller status? Was the original muscle car still the best? In April, *Car and Driver* pitted the Pontiac against its five main rivals, and was ambivalent about the result. "Certainly the sportiest looking and feeling car of the six," went the write-up, "Its shape, its paint, its flavor, say GO!… but it was its suspension that let it down." It suffered severe axle tramp, "bordering on the uncontrollable," and only the Comet was worse. The test driver for track sessions at Bridgehampton (racing driver Masten Gregory) was unimpressed by the GTO's

cornering power as well: "The suspension is certainly too soft … it tends to float and bounce in the corners."

It might still have won out as a comfortable boulevard cruiser with a strong image, except that the test car came with the Royal Bobcat tuning package. An option offered by a Michigan Pontiac dealer, this consisted of richer jets, thinner head gaskets, positive locking nuts, and retuned distributor. These were not huge changes, but they made the GTO the most powerful car of the six, despite having the smallest engine. The trouble was, it turned the quiet, docile 389 into a recalcitrant beast. Hard to start and reluctant to keep going when cold, it idled noisily at 1,000rpm and guzzled gas at 11mpg (5.5km/liter) even at steady speeds. To add insult to injury, the fuel pump drive sheered off during the test, the 389 shed its fan belt on the drag strip and the left upper control arm of the rear suspension broke. Twice the GTO had to be sent back for repairs before the test could continue. "When [the GTO] was first produced," concluded *Car and Driver*, "we tended to forgive some of its handling foibles because of its newness and exciting originality, but now, in the face of sophisticated packages like the 4-4-2, it needs improvement."

A Change of Climate

One generally associates the early 1970s with a downturn for muscle cars, when emission and safety standards were beginning to bite. But the climate had been changing from a time years earlier. California-bound GTOs had been fitted with closed crankcase ventilation from a time early in 1964, and with air injection from 1966. For 1967 the "GeeTO Tiger" was dropped, and even John DeLorean emphasized safety in his public statements.

As for the GTO, it acquired a whole range of new safety features: dual-circuit brakes with those long-overdue and badly-needed discs (still only an option though), soft window handles, and coat hooks, a breakaway mirror and a collapsible steering column. Just a few years' worth of muscular tire-smoking antics, with the emphasis firmly on performance, had been enough to have safety campaigners up in arms. If one considered the number of people killed on U.S. roads each year, the critics had a point, especially while manufacturers paid so little attention to safety. So maybe it is possible to liken the 1960s muscle car era to a teenage party: it was not yet over, but responsible parents were standing around outside the house, tut-tutting about the noise and bad behavior, and it would not be long before one of them marched in and pulled the plug out of the stereo.

So Pontiac soft-pedalled its advertising for 1967, which merely disguised the fact that alongside the new safety features was a bigger engine. With a new block, the 389 was bored out to a full 400 cu in (6.55 liters) to take advantage of the GM capacity limit for intermediates. There was actually an economy version of this engine, with low 8.6:1 compression and two-barrel carburettor, rated at 255hp (190kW), and this came with an auto transmission only; less than 4 per cent of GTO buyers actually chose it. One suspects it was a sop to the safety campaigners. Most buyers went for the standard four-barrel 400, for which Pontiac claimed the same 335hp as the old 389. In place of the Tri-Power, there was a 400HO (High Output) version with 288/302 cam, free-flowing exhaust manifolds and a claimed 360hp, again exactly the same as its predecessor. The ultimate GTO power unit for 1967 was the Ram Air, which used a working air scoop to funnel cold, dense air straight into the intake at speed. As part of the package, it came with a peakier camshaft (301/313 timing) and low rear axle ratio of 4.33:1. That made it a mite buzzy on the highway, and the tuned Ram Air could get hot and bothered in warm weather.

Despite all the extra equipment, this engine put out the same 360hp as the HO, if one believed Pontiac. But really this was a sign of the times: manufacturers

were understating power outputs to avoid drawing flak from the safety campaigners. Maybe they went too far, for only a small minority of GTO buyers paid extra for Ram Air in 1967. The actual figure was less than one in a hundred.

In the new climate, Pontiac would in any case have preferred to concentrate on the new disc brake option, which brought vented power-assisted front discs with four-piston callipers. There were separate circuits for the front and rear, so there were still brakes if one of the circuits failed. A three-speed automatic, the TH-400, was another new option, finally replacing the ageing two-speed. More noticeable was the extra-cost hood-mounted tachometer. It was a neat and novel idea, though in practice tended to mist up in damp weather and was difficult to read at night, while its delicate components did not cope well with full-blooded hood slams.

As with most muscle cars, options were part of the fun. The GTO buyer faced lists and lists of options: quite apart from the three body styles, 12 different engine/transmission combinations, and ten rear axle ratios, there were the endless detail touches available either on the production line, or dealer-fitted. The dealer could, for example, fit an engine block heater, ski rack, or litter basket (the last in red, blue, black, or beige). Factory options were extremely involved, with everything from the Ram Air engine to heavy-duty electrics or a reading lamp. If there seemed to be too much choice (there were nearly 100 individual factory-fitted options alone), there were various option groups. To give some idea of just how extensive this choice was, have a look at the tables. For the production planners, parts men, and dealers, it must have been a nightmare.

GTO sales fell in 1967, from that all-time high, to 81,722 vehicles. What with the new pro-safety climate and a horde of bigger, more powerful competitors, one might have forgiven Pontiac for expecting worse in 1968. But the GTO bounced back. With new more rounded styling and more power, sales crept back up to over 87,000, and the GTO was even voted Car of the Year by *Motor Trend*. (The new shape had much to do with this fresh lease of life.) Now in hardtop and convertible forms only, the GTO was more sporting than before, the classic long-hood/short trunk shape on a shorter 112-in (2.84-m) wheelbase. This was shared with the Tempest, but the GTO added trimmings like out-of-sight wipers and dual exhaust tailpipes. More obvious was the chrome-free Endura rubber bumper. Made from high-density urethane-elastomer, it would resist denting and even bounce back into shape after minor knocks. However, the standard chrome version was also available. The concealed headlamps were a new option, echoing the 1967 Camaro, and revealed by slide-away vacuum-operated doors in the grille. That was fine, until a leak developed in the actuator, and "winking" GTOs (one headlamp door closed and the other open) were often seen as the cars aged.

There was an all-new interior as well, with lots of wood grain trim and a new three-pod instrument panel, though on this supposedly sporting car one still had to pay extra for a tachometer. Despite new moves to clean up emissions (redesigned combustion chambers and changed ignition timing) the 400-cu in (6.55-liter) V8 offered more power, with 350hp (261kW) in standard four-barrel form or 265hp (197.5kW) in two-barrel form. Meanwhile, a new Ram Air II system featured cylinder heads with bigger intake ports and freer-flowing exhaust ports, a cam with hotter timing and higher lift, plus bigger valves. Power was up, but only slightly, to 366hp (273kW), while the non-Ram Air 400HO remained at 360hp. There was just one drawback to Ram Air: it could not be used in the rain, as the open hood scoops let water in. So it came as a package in the trunk of the car; at the first sign of rain, the driver had to stop and swap back to blanked-off scoops. It is hardly surprising that Pontiac did not sell many.

The big news for 1969 was The Judge. Pontiac had long been pondering the fact that the GTO range fell short of both a budget muscle car and a premium model,

OPPOSITE, ABOVE & OVERLEAF:
The Judge was really a GTO with bright
paint and eye-catching graphics.
Mechanically, it used the 400HO with
Ram Air, plus heavy-duty suspension, and
wide, low-profile tires.

during a period in which the competition
offered both. There was a whole range of
budget performers, such as Pontiac's own
350 HO-powered Tempest, which were
cheaper than any GTO and cost less to
insure, an important factor for younger
drivers. But it was never promoted by
Pontiac as a muscle car, and sales
reflected the fact. So company engineers

put a proposal together for a hopped-up Tempest to fill the gap and sent it up to John DeLorean. He rejected it, and ordered that the car be upgraded as a new range-topper for the GTO instead. This was The Judge.

It was, if truth be told, little more than an existing GTO with Ram Air and loud colors. Pop art was influencing car design in the U.S.A., and bright colors with suitable decals were becoming part of the muscle car look. Sure enough, the first 2,000 Judges were finished in Carousel Red, with extrovert 60-in (1.52-m) rear spoiler, bubble-letter "The Judge" logos, and stripes. Mechanically, the Judge used the 1968 400HO power unit, but with Ram Air III, heavy duty suspension, Rally II wheels, and G70 x 14 tires. For an extra $332 over the price of the standard GTO, it offered a real eyeful, and came in both hardtop and convertible forms, though only 108 of the latter were built, compared with nearly 7,000 hardtops.

There were few other major changes for the GTO that year, though the latest Ram Air IV system now claimed 370hp (276kW) at 5,500rpm as a result of improved intake ports and limited-travel hydraulic lifters, plus a slightly higher lift on the cam. Both this and Ram Air III, incidentally, were now driver-controlled. If it rained, the driver not longer had to stop, get   out, find those closed-off hood scoops and fit them. Now, the driver simply pulled a knob and the scoops closed themselves.

Seventies Slide

It was just as well that Pontiac had the Judge for 1969, and sold nearly 7,000 of them, for GTO sales slid to a little over 72,000 that year. That was a drop of almost 20 percent, and heralded the rapid slump in GTO sales during the early 1970s. In 1970 they fell by nearly half, to 40,000, and to just over 10,000 in 1971, when the GTO was outsold ten to one by its bread-and-butter stable mate, the Tempest. Its fall from grace was sudden and dramatic, and its end was near. In a way, the GTO's tough performance image was biting back: along with other muscle cars in the early 1970s, it was suffering from spiralling insurance costs, especially for young drivers, as experience revealed that cheap tire-smoking muscle cars and the under-25s were not always a good mix.

However, if the 1970 GTOs were anything to go by, it was simply business as usual. There had been complaints that the 1968 bodyshell left the GTO looking a little too refined and effete. So the 1970 version was given butch wheel-arch moldings and a new aggressive front end with quad headlamps. There was a similar theme under the hood. GM had belatedly lifted its 400-cu in capacity limit, and Pontiac responded with a 455-cu in (7.46-liter) V8. However, this was not the high-horsepower option. Instead, the 455 was a relaxed, torquey engine in a mild state of tune, aiming to provide a suitable boulevard power unit for luxury GTOs: that was underlined by 360hp (268.5kW) but over 500lb ft (678Nm) of torque. Otherwise, the familiar 400s continued, offering from 350hp in the standard four-barrel to 370hp with Ram Air IV.

Incredibly, for 1970, 9.5-in (24.1-cm) drum brakes all round, with no power assistance, were still the standard brake set-up, though handling was at last improved with a rear anti-roll bar and a larger front bar. Spring rates were unchanged, but the suspension mountings were beefed up and a new variable-ratio power steering system was offered. The result was a great improvement, and the GTO finally had handling to match its straight-line performance. The Judge was still part of the line-up ("After a few moments of respectful silence," went the 1970 advertisement, "you may turn the page") and there was a new GTO-related budget muscle car, the GT37, based on a Tempest two-door. And there was an unusual new option that year. The Vacuum Operated Exhaust allowed the driver to move the

ABOVE & OPPOSITE: Most Judges were coupés. The V8 under the hood was identical to that of the GTO.

exhaust baffles via an dashboard switch, giving more noise and horsepower at the pull of a dashboard knob. It was not quite as anti-social as it sounds, with the choice between a fully-legal standard GTO exhaust and the quiet LeMans equivalent. Only 223 GTO buyers specified the VOE.

A GTO Judge, drag racing.

By 1971, Pontiac seemed to have given up on the GTO. It was changed very little, apart from a bigger grille and hood scoops, and there was no advertising in the mainstream magazines at all. Compression ratios were dropped to allow all GTOs to run on regular fuel, and a new round-port version of the 455

HO was listed at just 335hp (250kW). It was clear that Pontiac had better things on its mind than promoting the ageing GTO: the Trans Am was now its performance flagship, while the Firebird had been redesigned for 1971. Ram Air III and IV were dropped. Power started at 300hp for the standard 400 (now with an 8.2:1 compression), with the 335-hp 455 the hottest GTO one could buy, though the car looked as macho as ever.

Only 10,532 GTOs found buyers in 1971, and for 1972 Pontiac did the obvious thing, and relegated it to option status: the GTO was now an option pack on the LeMans, though that did make it cheaper. Of course, this was exactly how the GTO had started, as a low-key option on a bread-and-butter model in 1964, which now seemed a lifetime away. Unless one was a GTO enthusiast, there seemed little point in buying one, as a LeMans could be specified at the same level of performance with a different range of options. Not surprisingly, GTO sales slumped again, to just 5,807 in that year. Of course, all muscle cars were suffering in the early 1970s, but the GTO lost out more than most. There were very few changes to the car, though what looked like a drastic power cut was due to a change to net rather than gross power measurement. The base 400 now started at 250hp (186kW), with 300hp (224kW) from the 455 HO.

The year 1973, according to author Thomas DeMauro, was a lost opportunity for the GTO. Those 20-something baby boomers who had bought the original were now approaching comfortable middle-age: they wanted more comfort and prestige, maybe even four doors, and DeMauro reckons Pontiac had just the GTO for them. Instead, it was unveiled in 1973 as the Grand Am, which sold well. According to DeMauro, Pontiac wanted the GTO to return to its roots as a proper high-performance muscle car.

So the 1973 model was based on the latest LeMans coupé with the controversial Colonnade styling. Not everyone liked it: the single front headlamps and heavy

chrome bumper seemed like retrograde steps, and it was certainly lacking the sheer threatening presence of the 1971 and 1972 GTOs. Early literature indicated that the car would be available with a 310-hp (231-kW) Super Duty 455, but when production started that was reserved for the Formula and Trans Am. Once again, the GTO option was not advertised, and even Pontiac's 1973 new model announcement contained just one, very small, reference to it. Only 4,806 people chose to add the GTO option to their LeMans in 1973, the worst sales figure ever.

By rights, the GTO should have died then and there. But at the eleventh hour Pontiac decided to give it another chance. This time they really did go back to the GTO's roots, reverting to the compact Ventura. Now this was promising. The Ventura may have been an economy car, almost identical to other X-body GM siblings like the Chevrolet Nova, but it was up to 800lb (363kg) lighter than the LeMans. Even with a smaller engine, it should in theory have provided reasonable performance with better fuel efficiency. The 1974 GTO, available as an option on both the Ventura and Ventura Custom in two-door or hatchback forms, was more than just a badge and paint job. It was fitted with its own 350-cu in (5.735-liter) V8, which had actually been around since 1968, but there were some detail changes to suit this latest application. With Quadrajet carburettor and mild cam timing, it produced 200hp (149kW) at 4,400rpm. This was not in the 400-cu in class, but it did have that vibrating air scoop bulging out of the hood. The rest of the car betrayed its economy origins, with the bland bar speedometer only slightly offset by a little engine-turned appliqué, a reference to the 1964 original. Performance-wise, the Ventura GTO did not live up to its promise of a good performance/economy balance: road tests indicated a standing quarter mile in around 16 seconds, but only 12mpg (5.5 liter/km). Still, some people liked it, and just over 7,000 GTOs were sold in 1974, a great improvement on the LeMans-based car of the previous year. It was not enough for Pontiac though, and the Ventura GTO experiment was not repeated in 1975.

But they say breeding will out. It did not take long for hot-car fanatics to realize that a 400 or 455 V8 could be slotted straight into the Ventura/GTO, to the point where original 350s are now hard to find. This equals a smallish car plus biggish engine, and that is where we came in.

The Pontiac GTO was relaunched in the United States in 2004, based on the Holden Monaro's V platform.

GM had high expectations to sell 18,000 units, but the Monaro-based GTO received a lukewarm reception in the U.S. The styling was frequently derided by critics as being too "conservative" and "anonymous" to befit either the GTO heritage or the current car's performance. In addition, the GTO faithful felt further insulted by GM's failure to present a U.S. built car that incorporated any design lineage from the muscular icons of the 1960s and 1970s. Given the newly revived muscle car climate, it was also overshadowed by the Chrysler 300, the Dodge Charger, Dodge Magnum, and the new Ford Mustang, which all featured more traditional "muscle" aesthetics. Sales were also limited because of dealer tactics, such as initially charging large markups and denying requests for test drives of the vehicle. By the end of the year, the 2004 vehicles were selling with significant discounts. Sales were 13,569 of 15,728 cars for 2004.

There were more modifications in 2005 but the car ceased production in 2006 marking the end of GTO's history.

FORD MUSTANG

It is likely that General Motors, Chrysler, and AMC executives assured each other that it would never work when Ford launched its Mustang during April 1964. After all, the Mustang was neither one thing nor the other. It was American-made, but without the comfort or space of a full-sized Detroit sedan. It was styled like a sports car, but had none of the nimble responsiveness of an MG or Porsche. In any case, who in their right minds would pay extra for what was basically a run-of-the-mill Ford Falcon, but with less room for people and shopping? They were convinced that the traditional American car buyer would not like it (it's too small and looks too radical), neither would the sports car fanatics (it's too big for those guys), or so hoped the rival manufacturers.

An early 1966 Mustang.

OPPOSITE: Ford Shelby Mustang GT350.

How wrong they were. Twelve months later, the news arrived that Ford had sold an incredible 419,000 Mustangs in its first year, a new first-year sales record for Detroit. Less than two years after the launch, the millionth Mustang was completed, and if anyone still harbored doubts about the car by that time, they were either out of touch or out of their minds. Not that any of Ford's domestic rivals could be accused of that: the moment the Mustang's massive success became clear, all of them began work on pony cars of their own, and the Chevrolet Camaro, Pontiac Firebird, AMC Javelin, and reborn Plymouth Barracuda were all directly inspired. All sought to tap into that youth-oriented market, whose existence Ford had so amply proved. But all were at least two years behind Ford, which left the Mustang with the entire happy hunting ground to itself, at least for a while.

The secret of the Mustang's success was simple. It was a reflection of the fact that mainstream American car buyers might actually want a little sports car glamour. For decades, U.S. car makers had given buyers what they thought they wanted: two- and four-door saloons of ever-increasing weight, power, and gaudiness. Though still heavyweights by European standards, more compact saloons had been built more recently, such as Ford's own Falcon in 1960, and the Chevrolet Corvair. But these were merely smaller versions of the same thing. The Mustang, with its long hood and short trunk, offered the sports car image at something close to saloon-car price. It was able to do this by taking most of its components from the Falcon, thus keeping costs down. Also, it was reassuring for buyers to know that under that Italianate styling lay reassuringly familiar components: straight six and V8 engines which everyone (including their parents) had been driving for years.

Another Mustang strength was its breadth of appeal. It may have been a youth-oriented car, but times were changing, and such products were not always confined to the young. "Within four months," wrote *Car and Driver* of the Mustang's early prophets of doom, "those oracles were watching 65-year-old retired druggists, school teachers, and just about the whole population of every semi-fashionable suburb in the country, standing in line to buy a Mustang." It was sporty and radical, but not over the top, and thus appealed to a wide spectrum of buyers. This was backed by a large range of engine and transmission options: by 1967 Ford was listing 13 different combinations. So the retired druggist could have a cheap, skinny-tired, three-speed manual straight-six, which was nice and docile for shopping, but was still different enough to cause a stir as he rolled up for the bridge game; but his 20-something grandson also bought a Mustang, if he could afford it, in the form of the latest 390 big-block GT, with four-speed, full instrumentation, fat tires, and fancy wheels. In short, the Mustang had created something completely new: the pony car.

Sense of Vision

It was all Lee Iacocca's idea. Countless others were involved, of course, and not everyone agrees as to just whose idea the Mustang originally was. Product planner Donald Frey and production expert Hal Sperlich, marketing man Donald Petersen and stylists Joe Oros and Dave Ash were all part of the Mustang project from an early stage. "That car was developed seven months before [Iacocca] saw it," said styling chief Gene Bordinat years later. "That car would have made it to the marketplace without Lee." But even if the idea for a smaller, sportier Ford had been around before Iacocca became involved, there can be no doubt that he was the power behind its transformation into something with massively wide appeal.

Looking back, it is easy to see big corporations like Ford as giant monoliths, led from the top (in this case by Henry Ford II). In reality, there was intense competition

OPPOSITE: The Mustang was based on this, the Ford Falcon. This is a Falcon Sprint.

ABOVE: 1960s Mustang convertible.

between senior executives to move higher up the greasy pole, with more intrigue and political maneuvring than in any medieval royal court. So gray suits on the way up (of which Iacocca was undoubtedly one) surrounded themselves with loyal acolytes: Iacocca's group called itself the "Fairlane Committee," and met early on Saturday mornings to discuss a new type of car they

called "small-sporty." At this point, Henry Ford II was not included in the deliberations.

The first clay model was for a sporty little two-seater, mid-engined and open-topped, with a gaping air scoop. It was shown to a group of sports car enthusiasts, who thought it was great. Iacocca later recalled: "I looked at the guys saying it – the offbeat crowd, the real buffs – and said, That's for sure not the car we want to build, because it can't be a volume car." His gut feeling was backed by Ford's finance

OPPOSITE: The 1966 Ford Shelby Mustang GT350 on the race track.

ABOVE: The engine bay of a race-prepared Mustang: note the strengthening brace.

department, which thought the two-seater would sell a paltry 35,000 cars a year which, for a mass-producer like Ford, was not worth pressing the "go" button for on the production line.

Then Iacocca dealt his master stroke, ordering that two bucket seats be added in the back, transforming the pure sports car into sporty family transport. "Up until that point," Donald Frey later recalled, "we had been thinking two-seaters. But [Iacocca] was right; there was a much bigger market for a four-seater." Even the

conservative money men agreed that four seats would widen the Mustang's appeal, perhaps to 100,000 a year, they said, though Iacocca thought it would be about twice that figure. Better still, basing what was now a 2+2 coupé on the existing Falcon platform and engines meant it would cost relatively little to get the new Mustang into production: a $75 million investment for 100,000 sales a year was more like it. In the summer of 1962, the stylists got to work, and Dave Ash came up with the long hood/short trunk shape that would became familiar to several generations of American drivers. Clay mock-ups were complete by August, and the water was tested with the Mustang II show car in October 1963. The models looked good, but there were still doubters.

Not least among them was Henry Ford himself, who had never warmed to this idea of a small, sporty Ford. Eventually, Iacocca persuaded him to come back to the design studio for one last look. This was make or break time: Iacocca had been preparing the ground for months, dropping hints throughout the company, even to the motoring press, about the little car's huge potential. "I'm tired of hearing about this goddam car," Ford is reputed to have said at that meeting. "Can you sell the goddam thing?" Iacocca assured him that he could. "Well you'd better." It may not have been enthusiastic, but it amounted to a "yes." The Mustang was on.

"The best thing to have come out of Dearborn since the 1932 V8 Model B roadster," declared *Car and Driver* when the Mustang was finally unveiled to the public and press on 17 April 1964. Gene Booth of *Road & Track* hit the nail on the head when he described the Mustang as "a car for the enthusiast who may be a family man, but likes his transportation to be more sporting." There were in fact two Mustangs at the very beginning, the notchback coupé and an open-top convertible, both with that big range of engines, transmissions and other options.

The basic powerplant was mild indeed, by later Mustang standards, in the form of a 170-cu in (2.785-litre) straight-six producing 101hp (75kW), though a few months later it was replaced by a sturdier seven-bearing version, of 200 cu in (3.28 litres) and 120hp (89.5kW). The V8 route started out with a 260-cu in (4.26-liter) 164-hp (122-kW) unit straight out of the Falcon. That too, was soon superseded by a gruntier version, in this case a 200-hp (149-kW) unit of 289 cu in (4.74 liters). At the same time, a higher-compression 289-cu in unit with a four-barrel carburettor was added, offering 210hp (156.5kW). But even that was not the most potent Mustang available during the first year. Three months after the initial launch, buyers could order the Hi-Po (High Performance) 289 Mustang, now with special cylinder heads, 10.5:1 compression, high-lift cam with solid lifters, 600-cfm (16.99-m_/minute) Autolite four-barrel carburettor and freer-flowing exhaust manifold. To cope with the extra power, the main bearing caps were beefed up, and the quoted figure was 271hp (202kW). With the optional low-ratio 4.1:1 rear axle, that was enough for a 14-second quarter-mile, according to *Car and Driver*, with a terminal speed of 100mph (161km/h). Predictably, Hi-Po was the favorite of the motoring press, in that it backed up the Mustang's sporty looks with serious performance: "... the HP Mustang backs up its looks in spades," said *Car Life*, while *Road & Track* summed it up as a "four-passenger Cobra."

But it was not just engines that forced new Mustang buyers to make choices. In place of the standard three-speed manual gearbox, they could have a four-speed manual or Cruise-O-Matic automatic three-speeder. And that was just the start. Assuming that one could afford them, power steering or brakes could be specified, as well as air conditioning and heavy-duty suspension. There were 14-in (35.6-cm) five-spoke steel wheels, a vinyl roof or wire wheel-style hub caps. One of the most

popular was the "Rally Pac," which was a steering column-mounted tachometer and clock, and proved popular because all first-year Mustangs came with the Falcon's standard instrument panel, which was anything but sporty.

In fact, there were so many options, both dealer and factory-fitted, that it seemed as if virtually every Mustang was different. It was the first "personal car," which buyers could "tailor" to their own requirements or ego. The reality was that with over 400,000 Mustangs sold in the first year, many would be identical, but the customers thought differently. One other point was crucial to the Mustang's feel-good factor: all of the cars, even the cheapest six, had bucket seats, a floor shifter, and sporty three-

OPPOSITE, ABOVE, RIGHT &
OVERLEAF:
A 1968 Mustang GT. Details counted on
this model: note the grille on the fastback.

spoke steering wheel, three items guaranteed to make an immediate (and favorable) impact in the showroom. To American car buyers in the early 1960s, these were icons of sporty, upmarket cars. They might have added a little to the cost of each and every Mustang, but first impressions count, and they were well worth having.

The Mustang made so huge an impression in its first few months that one could have forgiven Ford for leaving the design as it was, and concentrating on churning out units as fast as possible for, after all, it could sell every single one. But Ford had rested on its laurels once too often, and did not relish a repeat. So three months into production came that Hi-Po V8; three months after that a third body shape, the "2+2" fastback, appeared, complete with fold-down rear seat: it complemented the long hood/short trunk look perfectly, and over 77,000 were sold in 1965. That figure made it a specialist model compared to the coupé and open-top Mustangs, over 600,000 of which were sold in that year.

The "Rally Pac" was all very well, but it soon became clear that Mustang buyers wanted something extra to differentiate the cockpit of their "small-sporty" from that of a next-door neighbor's Falcon. In April 1965 they got it. The "interior décor group" soon became known as the "pony interior," thanks to the galloping horses embossed on all four seats. It may seem rather quaint now, but then it was cool, and that was not all that the buyer got. As part of the package, there was simulated walnut for the steering-wheel rim and various other bits and pieces, door-mounted courtesy lights, various bright moldings and "pistol grip" door handles. Best of all, that boring Falcon instrument strip was replaced by a proper sports car set-up with five round dials.

That took care of the inside. At the same time, Ford introduced the GT option, which brought power-assisted front disc brakes, a quicker steering ratio (22:1 rather

than 27:1) and stiffer springs, bigger shocks and front stabilizer bar. And in case bystanders were in any doubt, the GT package also included look-at-me parts: body stripes, twin tailpipes, and fog lamps. Only the powered-up 225-hp and 271-hp V8s could be had with the GT pack, so there was no disguising your bargain basement six. If one were really strapped for cash, the Sprint 200 of 1966 offered a flashier version of the basic 200-hp Mustang, with pinstripes, wire wheel covers, center console, and chrome air cleaner. It was all very well, but in the meantime there were new kids on the block. With the new Camaro and Firebird imminent, the Mustang would not be having its own way for much longer.

Meeting the Challenge

In 1967, the Mustang changed, as part of its first major facelift becoming longer, wider, and slightly taller. It put on 130lb (59kg) in weight and the new styling reflected this, making it slightly more beefy and bulbous, as if the lithe young contender of 1964 was starting to acquire middle-aged spread. Not everyone appreciated the change, especially Iacocca, who saw the upsized Mustang as a dilution of his original concept. It was, he later wrote in his autobiography, "no longer a sleek horse, it was more like a fat pig."

But there was a need for change. For its first two and a half years, the Mustang had the pony car market to itself: it had, after all, created the niche in the first place. But in late 1966 serious competition appeared in the forms of the Chevrolet Camaro and Pontiac Firebird. GM's own pony cars were aimed right at the Mustang's jugular: the styling may have been a little conservative by comparison, but there was a crucial difference. From the start, the Camaro came with a big-block 396 V8 and the Firebird a 400, and either of these could blow even the hottest Mustang GT into the proverbial weeds.

For a car with sporting pretensions, this was embarrassing, so the priority was to give the Mk II Mustang some extra horsepower. That was partly why it had to grow, as Ford's big-block V8 would not otherwise fit under the hood. The engine they chose was the FE-series, in 390-cu in (6.39-liter) form, which put out 320hp (238.5kW). The former range-topping Hi-Po 289 was still available, though it would soon disappear, and both engines came with or without the GT option package.

It looked good on paper, but was not quite enough to keep up with the opposition. Early in 1968, *Car and Driver* tested a Mustang 390 GT against five rivals: every major U.S. manufacturer was now in on the pony car act, so Ford had an even tougher job on its hands. Running a 14.8-second quarter made the Mustang slightly quicker than the Camaro 396SS, but C&D reckoned it to be "a lame gazelle in the street racers' sweep stakes." The test car was not helped by a substandard engine that was recalcitrant from cold and needed a tune-up before it was finally up to speed. Even with a 100 per cent healthy power unit, the Mustang would not have fared well. Testers criticized its tendency to understeer and very heavy steering. Nor did they like the 2+2 fastback shape, which was exceeding awkward for rear seat passengers and difficult to see out of, with poor luggage room. On a points basis, the Mustang 390 scored lower than any other car. In short, it had gone from toast of the town to turkey in just three years.

In its conclusion, however, *Car and Driver* admitted that although the Mustang scored lowest, none of the testers considered it the worst car, so it must have had some redeeming features. The American public seemed to agree, and continue to

buy the Mustang in large numbers: the Camaro did not finally overtake it until the early 1970s. As ever, there was huge scope for personalizing the Mustang: 13 engine/transmission permutations alone, thanks to the new big-block V8, plus new options like a sun roof, cruise control, and integrated air conditioning. However, these luxury touches suggested that the Mustang was going a bit soft, forcing the true performance enthusiasts to look elsewhere. Late in 1967, one Ford dealer revealed that 390 sales, healthy at first, had dwindled "... to practically nothing. We found the car so non-competitive," he went on, "we began to feel we were cheating the customer."

Looking to the future, it was crucial for Ford to win the performance crown back from GM, though the lead was actually taken by Ford dealer and tuner Bob Tasca, who blamed the 390's lack of

Mustang GT with equipment.

"pizzazz" for falling sales among the performance buffs. To regain this factor, Tasca used no expensive, hand-crafted tuning parts, just regular off-the-shelf Ford components that happened to perform better than the standard 390. Tasca removed the engine from one 390, replacing it with a 428 Police Interceptor unit with big-valve heads and a 735-cfm (20.81-m_/minute) four-barrel Holley carburettor. The result transformed the Mustang into a real hot-rod, able to run the standing quarter in 13.39 seconds. They nicknamed it KR for "King of the Road", and it was.

Ford very quickly realized the potential and introduced its own production version, the 428 Cobra Jet Mustang, during April 1968. This was not a mere existing engine shoehorned under the bonnet. Instead, Ford mixed and matched parts to create a 335-hp (250-kW) V8

GT badges, wood-rimmed wheel and five-dial instruments were part of the GT package.

that brought it back to the top of the performance pile. The Cobra Jet unit was based on the existing 428-cu in (7.0-liter) V8, but with high-compression 10.6:1 pistons, a 390 GT cam, Holley four-barrel carburettor, a free-flowing exhaust and Ram Air. To cope with a claimed 335hp, the CJ Mustang came with power-assisted front discs, braced shock towers, staggered rear shocks, and limited-slip differential. *Hot Rod* magazine was delighted, summing it up as "probably the fastest regular production saloon ever built."

In fact, the late 1960s would be a real performance-fest as far as the Mustang was concerned. Much of this was down to Semon E. "Bunkie" Knudsen, who joined Ford from General Motors in February 1968. He loved performance cars, and made no bones about his ambitions for Ford: to become top dog in that particular market. Knudsen was also president of Ford, so the next 18 months saw three high-horsepower Mustangs that were faster than anything seen before.

In a way, these three were the Mustangs closest to the muscle car tradition, but each one was carefully targeted at a different market: the Mach 1 was a relatively civilized road car; the Boss 302 was aimed at handling prowess, and the Boss 429 was a hairy, straight-line machine pure and simple. Now with new smoother styling for 1969, the Mach 1 SportsRoof (the latest name for the fastback) was powered by a small-block 351-cu in (5.75-liter) V8 in a fairly mild state of tune. But to do justice to its name, hood scoop and go-faster stripes, buyers could specify a hotter four-barrel 351, the 320-hp (238.5-kW) 390 GT or of course the 335-hp 428 Cobra Jet. But even with the standard 351, Mach 1 buyers could enjoy wide E70 tires, "Special Handling" suspension and interior upgrades like high-backed bucket seats and a center console. The Mach 1 was a success, with well over 70,000 sold in its first year.

The Boss Mustangs were very different, being sold in far fewer numbers, and were really aimed at the race track and drag strip, though of course they were also road-legal. Knudsen had ordered a new Mustang that would be "absolutely the best-handling street car available on the American market." So the Boss 302 sat on lowered suspension and wide F60 tires on 7-in (17.8-cm) rim wheels, with power from a high-revving small-block V8, the 302-cu in (4.95-liter) with Ford's free-breathing Cleveland heads. Surprise, surprise, it also got the Mustang conveniently inside the 305-cu in (5-liter) capacity limit for Trans Am racing. Less than 2,000 people ordered Boss 302s in 1969, but the stripes, spoilers, and bold colors ensured that all of it would be noticed.

The Boss 429 was different again, relying on the brute force of a big, heavy engine rather than a smaller higher-revving one. So big was the engine, indeed, that the front shock towers had to be moved apart by 1 inch (2.54cm) to make room. With 375hp (279.5kW), the Boss 429 was without doubt the most powerful factory Cobra yet, though its 13.5-second quarter-mile was no faster than that of the original Tasca 428. The truth was that the biggest Mustang was no more than a convenient means of homologating the 429 "semi-hemi" (its hemispherical combustion chambers owed much to the Chrysler Hemi) for NASCAR racing. Regulations did not specify the car in which the engine was used, only that at least 500 be sold, and in this case it happened to be the Mustang. In any case, the ploy worked, as 859 Boss 429s found buyers in 1969: peanuts really, but worth it for the media exposure.

The big-selling Mustangs were cheaper, more practical cars like the Mach 1 and the new Grande hardtop offered in the same year. This was a luxury Mustang, with softer suspension, deluxe interior, wire wheel covers, more sound insulation, and a vinyl roof. Over 20,000 were sold in 1969, and nearly 13,600 in the following year. Still, those figures underlined the fact that in overall terms sales of the Mustang were down. Although still the best-seller in its class, the Mustang now shared its cake with five hungry rivals. Who would win?

OPPOSITE, ABOVE & OVERLEAF:
Mustang Boss 351, with an all-black
interior and SportsRoof Form.

Serious Downsizing

The early part of the 1970s was not a
good time to be making muscle cars. For
a decade, manufacturers had been
happily pursuing a horsepower war, each
year bringing ever more tire-melting, fuel-

OPPOSITE & BELOW:
Mustang Boss 351.

guzzling monsters onto the roads. But now the new challenges were not in the manufacture of the most powerful car on the market, but keeping up with ever tighter safety and emissions legislation, while insurance premiums hit the roof. Bunkie Knudsen did not foresee all of this (though as president of Ford maybe he should have done so), so the new

Mustang that was authorized by him was the fattest, heaviest yet. Officially, Knudsen had been "let go" from Ford, and after a mere 18 months at the top. The real reason was that there was insufficient room for both Knudsen and Iacocca at the top of the same company, and while Iacocca was a Ford man through and through, Knudsen was an outsider. Henry Ford had once said that history is bunk. "Bunkie is now history," was the gleeful adaptation throughout Ford after this ex-GM man packed his bags in September 1969.

But before he left, Knudsen had given the nod to a new Mustang. Just like the 1967 version, the Mustang for 1971 was longer and wider than before. It also weighed an extra 250lb (113kg), while the basic price for a V8 coupé jumped to over $3,000. If the 1967 Mk II Mustang had shown early signs of middle-age spread, then this one was heading for outright obesity. Predictably, Lee Iacocca was no more impressed by this latest Mustang than he had been by the 1967 "fat pig."

It was the power race that was behind it all: to keep up with the competition, the Mustang had to keep growing. Sure enough, 1971 saw the 429 Cobra Jet replace the old 428, using Ford's all-new 429-cu in V8, which replaced the FE unit. Claiming 370hp (276kW), it gave the new Mustang a 13.97-second quarter-mile time, according to *Super Stock* magazine. Complementing this was the latest SportsRoof, now with a radical 14-degree fastback: *Car Life* magazine thought "flatback" would be a better term, pointing out that the extreme angle reduced rear-view vision to almost nothing. Not that there was anything new in that, as Mustang fastbacks had always sacrificed practicality to looks.

But whatever bodywork they wore, the days of the old-style heavyweight muscle cars were numbered, and less than 2,000 Cobra Jets were sold in 1971. There was also a spiritual successor to the Boss 302, the Boss 351 from November 1970. It carried on the line of high-revving, small-block V8s, in this case the 351-cu in (5.75-liter) 351 HO with ultra-high 11.7:1 compression, 715-cfm (20.25-m_/minute) four-barrel carburettor and 330hp (246kW). To back that up as part of the Boss package, there was a Hurst four-speed shifter, competition suspension and F60x15 tires, with full instrumentation inside. The press liked it: *Motor Trend* reckoned it was as quick as the big 429 Cobra Jet, and in truth the Boss 351 was one of the fastest production Fords of all.

However, times were changing, and the market was adopting new priorities so different from the free-living, high-horsepower, high-octane 1960s. That was reflected in falling Mustang sales, by 22 per cent in 1971 alone, and it was clear that Knudsen's timing could not have been worse as the biggest Mustang was already starting to look like a dinosaur. Reflecting that feeling, the basic power unit for the Mach 1 was downgraded from the 351 to a 302-cu in (4.95-liter) small-block. A sign of the times, sales fell again in 1972, and though they recovered slightly in the following year, the changing market was underlined by that luxurious Grande. As the fat-tired sporty Mustangs sat in showrooms, Grande sales climbed 28 per cent in 1971, and by 1973 made up one in five Mustang sales.

Iacocca got his revenge on Knudsen. In late 1970 he was appointed president of Ford. Now it was payback time for the Mustang humiliations of 1967 and 1971, which in Iacocca's view were mere travesties of the original concept. Sales between 1971 and 1973 seemed to bear him out. Iacocca wanted to return the Mustang to its roots as a sporty-looking, easy-to-drive car, cheap to buy, and economical to run, and above all, small. Iacocca was not talking European small here, but when it was unveiled in 1974, the Mustang II looked like a true pony car next to its predecessor. It was 4in (10.2cm) narrower, over 1ft (30.5cm) shorter in the wheelbase and with

300lb (136kg) trimmed off the kerb weight. Previous Mustangs had kept a family resemblance to the 1964 original, but this machine, styled by Ghia of Italy, was all-new. However, like its grandfather, the Mustang II was based on a small bread-and-butter Ford, the Pinto.

By no stretch of the imagination could it be described as a muscle car. Purists must have gazed, horrified, at the press reports. The base engine unit was a mere 140-cu in (2.3-liter) four, with a 171-cu in (2.8-liter) V6 optional and, worse still, there was no V8 option at all! However, it did fit the mood of the times, and as the

OPPOSITE, BELOW, OVERLEAF & 82–83:
The Mustang Mach 1 428 Cobra Jet.

Ford ads said: "The right car at the right time." Well, maybe it was in the short-term. In its first year, the Mustang II recreated that original 1964 stampede, with almost 386,000 sold. Would a rip-snorting V8 have sold that many in the dark days of fuel crisis?

"Scoops aplenty in this 428 Cobra Jet – no one ever said muscle cars had to be subtle."

The honeymoon was short-lived, however, and as gas prices began to ease down again, so did sales of the economical Mustang II. They slumped by half in 1975, then more gradually in 1976 and 1977, though actually bounced back again in the car's final year, up by 25 percent. Reflecting the

OPPOSITE & ABOVE: The 428 brought the Mustang back into contention as a performance car.

times, Ford introduced a V8 option in 1975, albeit a mild-mannered 302, and butched-up the Mustang II with the striped and spoilered Cobra II in 1976 and low-production King Cobra two years later. Power front discs, power steering and heavy-duty suspension were part of the King Cobra package, as was a four-speed manual gearbox. But its 302 V8 was a pale shadow of its fire-breathing predecessors, with just 139hp (103.5kW) at 3,600rpm. Scorned by a generation of enthusiasts since, the Mustang II deserved better: it sold in greater numbers than its immediate predecessors, and kept the name alive until performance was back in fashion.

A New Generation

At first, the third-generation Mustang of 1979 seemed like more of the same. The project had begun back in 1973, with Ford already looking forward to what would replace Mustang II: the company wanted a light, fuel-efficient coupé, and that is what (by American standards) it got.

Like the Mustang II and the original 1964 model, the newest Mustang was based on an ordinary platform, in this case the Fairmont saloon, which had been launched in August 1977. It was wider, longer, and taller than the car it replaced, had 20 per cent more room inside, but was 200lb (91kg) lighter. The top and bottom engine options were unchanged, in the forms of a 139-hp 302 V8 and 140.4-cu in (2.3-liter) four respectively, but the big news was a turbocharged version of the four delivering around 130hp (97kW), indicating that Ford was serious about offering performance again, but in a more efficient package. Suspension was by MacPherson strut at the front, with four-link suspension at the back, in either standard or sports TRX form (with attractive alloy wheels as part of the package). At first, it came as a two-door notchback or hatchback only, though a convertible followed in 1983.

History repeated itself when the new Mustang went on sale in 1979, with almost 370,000 cars finding homes. Just like the first Mustang in 1964, and the second ten years later, the first year was the most successful. The new Mustang was given another boost when it was chosen as official pace car for the Indianapolis 500, just as the original had done 15 years earlier. Ford made the most of that by selling 11,000 Indy pace replicas, with either V8 or turbo four power. Neither of these power units did much to excite true muscle car freaks but, over the next few years, the Mustang would gradually creep back into the performance market, recovering the crown from the Camaro.

The first sign that the Mustang would not remain meek and mild forever came less than two years after its launch. The McLaren Mustang was a low-production product of Ford's new Special Vehicle Operations (SVO) department. Its real purpose was to act as a showcase for what SVO could do, and stimulate existing Mustang owners' interest in aftermarket tuning parts. So only 250 were built (some say fewer than even that) and the price was a sky-high $25,000. For the money, one had a 175-hp (130.5-kW) version of the turbocharged four, with turbo boost turned up to 11lb/sq in (0.76 bar) from the figure of 5lb/sq in (0.345 bar) in the standard car. To avoid confusion with the standard production line Mustang, the McLaren carried a full body kit of flared fenders and big spoiler, with BBS wheels and 225/55 Firestone tires. There was even a roll cage!

It would be another four years before a production version of the McLaren appeared. Named after that specialist department, the Mustang SVO also squeezed

PREVIOUS PAGES, ABOVE & OPPOSITE:
Mustang Mach 1.

PAGES 94–95: Shelby's ultimate Mustang
was the GT500.

*"Bright color, matt black
stripes, spoilers and
louvres, all-black interior:
this was the late-
1960/early- '70s muscle
car personified."*

175hp out of the turbo four with the aid of fuel injection and an air-to-air intercooler. A large hood scoop delivered cold, dense air to the turbo, helping boost torque to 210lb ft (284.8Nm) at 3,000rpm, and a five-speed manual transmission was standard. The whole package was performance-orientated, with all-round disc brakes, a Traction-Lok rear axle and aluminium wheels with 7-in (17.8-cm) rims. Ford claimed a top speed of 134mph (215.5km/h) and a 0–60mph time of 7.5 seconds. If press reports were anything to go by, that was no overblown claim either. "The SVO outruns the Datsun 280ZX," said *Road & Track*, "outhandles the Ferrari 308 and Porsche 944, and it's affordable." It got a power boost in 1985, to 205hp (153kW), and nearly 10,000 SVO turbos were built in three years.

So did that mean the end of the V8 as the muscle Mustang's power unit of choice? The simple answer is, not a bit of it. The 302-cu in (4.95-liter) motor had been downgraded to a 255-cu in (4.18-liter) in 1980, but from 1882 onward power crept up and up. "The Boss is Back," announced the ads, when the full 4.95-liter returned in 1982, in 157-hp

(227kW) two-barrel form and with the High Output (HO) tag. The year after, a four-barrel Holley carburettor boosted that to 175hp. All Mustangs, incidentally, came in LX or GT trim.

The V8s were proving popular. This was not only because they remained first choice for U.S. gas heads, but also because they were substantially cheaper than the SVO turbo: in Mustang GT hatchback form, the latter was priced at $14,521, the V8 less than $10,000. With economics like that, and the oil crises of the 1970s fast becoming folk memories, the new breed of Mustang V8s could hardly fail. So for 1985 Ford introduced the latest 4.95-liter HO Mustang, now with 210hp (156.5kW) at 4,400rpm and 270lb ft (366.1Nm) at 3,200rpm. The extra power came from a high-lift cam, roller tappets and two-speed accessory drive. It was hardly surprising, therefore, that over 36,000 V8s were sold in 1984, and over 45,000 the year after. Less than 2,000 SVO turbos found buyers in 1985, making it clear that it was back to the good old days for American car buyers: one could still have a Mustang in basic 88-hp (65.5-kW) four-cylinder or V6 form, but the V8 was king.

So successful was the third-generation Mustang that Ford abandoned plans to replace it with a front-wheel-drive Mazda-based car (this became the Probe) and allowed the Mustang to carry on. There were fewer changes in the late 1980s, though the 4.95-liter HO V8 was now up to 225hp (168kW) at 4,200rpm and 300lb ft (406.8Nm) at 3,200rpm, thanks in part to sequential fuel injection. Fundamental to its appeal, as had been the case with the original Mustang, was performance per dollar. In 1989, V8 prices started at just $11,410 for the LX coupé, and a GT hatchback was $13,272, though convertibles were $17,000 or more. That led *Motor Trend*'s Automotive Yearbook to sum it up succinctly: "It's a very hot package with a 1960s ring and a torque curve that'll tighten your skin, all for the price of a Honda Accord."

So the 1980s Mustang had reinvented the good-value V8 muscle car, not long after many drivers thought those days had gone forever. It also enjoyed a very long production run, and was not replaced until 1994; but for its last year, Ford unveiled something special. The Mustang Cobra for 1993 was a product of Ford's Special Vehicle Team (SVT), the 1990s equivalent of the SVO. But it was not only an end-of-line boost: Ford needed to hot-up its now ageing Mustang in the face of the new 275-hp (205-kW) Chevrolet Camaro and Pontiac Firebird. Ford had actually considered building a 351-cu in (5.75-liter) Mustang in the previous year but rejected the concept as too costly. Instead, the familiar 302 was given cast-iron big-valve GT40 cylinder heads, a two-piece intake plenum and manifold, a peakier camshaft, higher-flow fuel injectors, and freer-flowing exhaust system, among other things. To cope with the extra power, the Borg-Warner gearbox was strengthened, there were all-round disc brakes and 7.5-in (19.05-cm) wide wheels. The suspension was actually softened a little: the art of suspension tuning had come on since the 1960s. The final 1993 Mustang was certainly quick, but not as fast as a Z28 Camaro, though that reflected its claimed 235hp (175kW) against the Chevrolet vehicle's 275hp (205kW); the 0–60mph and quarter-mile figures were 6.2 and 14.4 seconds respectively, while those of the Camaro were 5.6 and 14.0, according to a *Road & Track* head-to-head test. It was fast enough, however, and for lovers of traditional V8 rear-drive muscle cars, a good note on which to end.

Today's line up of Mustangs are as popular as ever. In particular, the new 2017 Mustang Shelby GT350R is all about performance and delivering immense power.

CHEVROLET CAMARO

General Motors had a problem, and it was called the Ford Mustang. Not that GM recognized the problem as such, at least at first. In fact, there seemed to be a consensus in Detroit that the Mustang would fail. Despite its massive advertising budget, the biggest ever in the industry, this was a mere Falcon with less practicality and radical styling. Who would buy it?

OPPOSITE & BELOW: The Camaro.
Was it the car to topple the Mustang?

The answer, as it turned out, was just about everyone: in the first year, over 400,000 Mustangs found buyers, and suddenly General Motors realized the magnitude of the problem after all, with Chevrolet taking it personally. Although Chevrolet constituted the budget end of GM, producing simple, straightforward, and fairly conservative cars, it was also

"Chevrolet's Mustang rival had to be faster than the Ford, roomier, better to drive, and cheaper. By and large it was all of those things."

OPPOSITE & BELOW: Camaro SS.

the sporty marque, and Ford's new baby was aimed directly at this sector. When the sales figures were published, and it was clear that the Mustang was more Model T than Edsel in its massive appeal, GM and Chevrolet had a rude awakening: they had to come up with a Mustang rival, and quickly.

The trouble was that both companies had been caught on the hop, like

everyone else in the industry, and it would be a couple of years before Mustang rivals finally began to hit the showrooms. But this very embarrassment seemed to give Chevrolet fresh impetus, seeing an opportunity to better the Mustang. So the new car would have to be better-looking than Ford's pony car, both inside and out; it would be longer, wider and lower with better legroom; faster and better-engineered, nicer to drive with better responses; and of course it had to be cheaper as well.

There had been a couple of similar proposals at Chevrolet already, one of which could have been on sale before the Mustang. The chief designer at Chevrolet, Irv Rybicki, had the notion of a smaller sporty car to rival the European imports, which by then were selling so well in the U.S.A. He was supported by Bill Mitchell, GM's overall styling vice president. But Chevrolet's general manager, Bunkie Knudsen, vetoed it: "the one thing we don't need right now," he said, "is another car." Knudsen did say yes to the Super Nova in 1964, however, a sporty show car designed to test public reaction; this time his boss Jack Gordon was the one to give the thumbs-down. Then those Mustang sales figures found their way on to top brass desks, and everything changed.

A design team under Hank Haga immediately set to work. The team knew what it wanted: not a small two-seater for sports car fanatics, or a hotted-up four-door sedan, but a two-door coupé, clean-looking, and with a clear Italian influence. They were also thinking of Chevrolet's own Corvair and a feature that had been there almost from the start: the hidden headlamps behind that egg-box front grille, a startling concept for 1966 and one that became the early Camaro's trademark. Standard on the RS and SS at launch, the headlights were revealed when that part of the grille pivoted sideways electrically, though there was a manual control should the electrics fail. Finally, like the Mustang, the new car would have that coke bottle kick-up over the rear wheels, suggestive of raw power.

But since the Camaro (though the name had not yet been chosen) was a feel-good car, the interior design was equally important. At the time, American-built cars were, from the driver's seat, uninspiring. Seated on the big bench seat, gazing at a bland strip speedometer, the driver could never feel special: he or she was really only one of the passengers. Chevrolet's designers realized that drivers in the Mustang class wanted something different. They needed to be flattered, made to feel important, as if they were in control of a precision, powerful machine: it was no coincidence that the Camaro ended up with several features influenced by fighter planes. Chief of interior design George Angersbach was so convinced of the wisdom of this that he had a four-speed manual gear shift built into his office chair, so that he could get into character! The result was an all-enveloping interior that wrapped around the driver: a triple pod of main instruments, with a central console stretching back between the front seats to carry more instruments, the radio, heater controls, and gear shift. It also placed a solid barrier between driver and passenger, so tight-lipped parents would probably approve as well.

Meanwhile, the engineering work was proceeding apace. There would be nothing radically new about the Camaro's mechanics, with drum brakes all round and a choice of existing Chevrolet engines, straight-sixes, or V8s. The new body was unitary-constructed, but like the Chevrolet II Nova (Chevrolet's compact saloon that bridged the gap between Corvair and Impala) also had a front ladder frame to support the engine and front suspension. The latter had independent coil springs and wishbones, while at the back monoplate leaf springs supported a solid rear axle.

Four existing engines were selected from the Chevrolet parts shelf: a 230-cu in (3.77-liter) six of 140hp (104kW), a 250-cu in (4.1-liter) six of 155hp (115.5kW), and a 327-cu in (5.36-liter) V8 in 210-hp (156.5-kW) and 275-hp (205-kW) forms, the latter aided by a four-barrel carburettor. But the Camaro would not have been convincing without a true performance option, so Chevrolet engineers produced a new 350-cu in (5.74-liter) 4bbl V8, producing a claimed 295hp (220kW). Even if few buyers actually opted for the projected high-performance Camaro SS, it would have done its job of letting its power/speed glamour rub off on the cheaper models.

BELOW & OPPOSITE
A 1969 and a 1968
Camaro SS.

OPPOSITE, ABOVE & OVERLEAF:
The 1968 Camaro had an optional 396-cu
inch (6.5-liter) V8, which was initially the
top-powered engine available.

After all, that was what most buyers in
the Mustang-class were laying out their
hard-earned dollars for – image.

This wide range of power options
naturally meant a wide range of running
gear to suit. Transmission options covered
a two-speed automatic as well as three-
and four-speed manuals; a five-speed
manual would not reach the Camaro
until the mid-1980s. Transmissions were

no problem, but wheels and tires were. From the start, the Camaro had been designed to look beefy and muscular, sitting on wide wheels and tires. That was fine for the powered-up V8 SS, with its D70 tires on 6-in (15.2-cm) rims, but the standard set-up for all other Camaros was 7.35 rubber on 5-in (12.7-cm) items. It looked, said Haga, "like a car on roller skates." Unfortunately, the designers and engineers would just have to live with it, for at General Motors the accounts department invariably had the last word. Whatever else it was or pretended to be, the Camaro had to make a profit.

Work proceeded swiftly. By mid-1965, prototypes were running around, heavily disguised, but recognizable as a new car: it was an open secret that Chevrolet was

working hard on a Mustang competitor. One of the few major problems to come up was that of severe scuttle shake on the convertible, though this was sorted out with counterweights. Road testing continued over the winter of 1965–66, and by the spring much of the work had been done and the Camaro was on target for its fall launch. But it did not have a name, and 5,000 contenders were considered. "Panther" and "Wildcat" were among the front-runners, a wild animal name having obvious attractions as a Mustang rival. But in the end, Chevrolet settled on Camaro because it sounded vaguely European and exotic: it also began with a C, which was a classic Chevrolet feature. But it was a fraught process, especially when a journalist discovered that a Spanish meaning of the word was a "shrimp-like creature." Then someone else found another meaning: "loose bowels." GM was to get a repeat experience many years later, when it discovered that the name of its new hatchback, Nova, translated into Spanish as "no go."

Names are not the most important feature of selling cars, however, and the bottom line was that the Camaro fulfilled its aims of outgunning the Mustang in a number of key areas: it handled better; all the various models were cheaper than their Mustang equivalents; and there was better trim and more options, like electric windows and folding seats. There were more safety features, too, such as twin-circuit brakes, collapsible steering column, and hazard warning lights. It was time for Chevrolet's copywriters to do their stuff, and they did not disappoint. "Meet the masked marvel," trumpeted one early advertisement. "Meet Camaro. Masked because it carries Rally Sport equipment with hideaway headlights. A marvel because it's an SS350: telltale domed hood, rally stripe and Camaro's biggest V8. Over 3200 pounds of driving machine nestled between four fat red-stripe tires, an SS350 carries the 295-horsepower 350-cubic-inch V8...Try one on at your Chevrolet dealer's. It's a ball and a half." Somehow, those unfortunate translations of its nametag were soon forgotten, and the Camaro was well-received when launched in late 1966. It was, wrote one road-tester, "a tasteful American interpretation of the European Gran Turismo." Haga and his boys were surely pleased with that one.

But it wasn't a clear-cut victory. In May 1967 *Motor Trend* magazine reported how the Camaro, in all its options, compared with the Mustang and Plymouth Barracuda. The magazine liked the wide range of engines on offer in the Chevrolet, including a new 396-cu in (6.5-liter) V8 option producing 325hp (242kW), not to mention the 400-hp (298-kW) Z28, there were seven. If the standard 396 did not have enough power, it could be hopped-up to 375hp (279.5kW) with a dealer-fitted camshaft; the Camaro was also the only one of the three to offer a larger-capacity (the 250-cu in/4.1-liter) six. With the 396 under the hood, the Chevrolet was decisively faster than the equivalent 320-hp (238.5-kW) Mustang as well: 0–60mph (95.5km/h) in 6.0 seconds, compared with the Mustang's figure of 7.4 seconds, and the standing quarter-mile (402m) in 14.5 seconds to reach 95mph (153km/h), as opposed to the Mustang's 15.6 seconds and 94mph (151km/h).

All three cars were available with stiffer suspension set-ups, but as standard there was not much to choose between them: the Camaro understeered a little less than both the Mustang and Barracuda, but was more prone to power oversteer than the Ford. But (and this was the one major criticism of early Camaros) it suffered severe rear axle hop under hard acceleration and braking. The optional suspension package (offered on the SS350 and 396) came with a traction bar that overcame this, and was thought to feel better than its rivals on tighter turns, the equal of the Mustang on higher-speed corners, despite the latter's heavy understeer. All three cars came with drum brakes as standard, and the Camaro's were judged the best, being without side pull.

The Camaro lost out to the Barracuda on interior space, however, and the Camaro convertible was judged to be more of a 2+2 than a full four-seater, just like the Mustang. The Barracuda won out on ride quality too, thanks to its longer wheelbase. Moreover, despite all the time George Angersbach and the interior designers had spent replicating a fighter plane's cockpit, the cheaper Camaros came with just a speedometer and fuel gauge. If one wanted extra instruments set into the central console, one had to pay extra, and one had also to choose between these

BELOW & OPPOSITE:
The Z28, initially a race-car option,
became a legend in
its own right.

and the stereo, as there was insufficient room for both. The bean counters had won out once again.

The Z28 is Born

But there was something else General Motors needed to out-Mustang the Mustang. It was not enough to be faster, better to handle, and cheaper. Since the Mustang class was all about image, it had to win races as well. This presented a problem, as GM had set its face against helping private race teams, let alone running its own. This was in accordance with the 1957 SCCA ruling, which banned factory-sponsored teams outright. However, Ford was flouting the rules by supplying plenty of support to the privateers, and if the truth be told, so was

RIGHT: General motors needed another plan to out-Mustang the Mustang. It was the Z28.

Chevrolet. But GM knew that to succeed in the new Trans Am series, a few warmed-over parts bolted into a standard Camaro would not be enough: what was required was a substantially new Camaro, designed specifically for racing. In making that decision, the company was on the threshhold of launching a legend, the Z28.

One can always tell when a car name is worth its weight in gold in marketing terms and image: it keeps coming back, year after year. So just as Pontiac's Trans Am tag was still in use up to the end of the 20th century, so were Ford's Mustang and Cobra. Over at Chevrolet, it was Z28, which made a comeback in 1977 and again in 2000. Legends die hard.

The substance of this particular legend was a new V8. Actually, it was not that new, merely a clever mix-and-match of existing parts. Trans Am racing had a 305-cu in (5.0-liter) capacity limit, and at first Chevrolet thought its existing 283-cu in (4.64-liter) V8 would neatly fit the bill. But it soon became clear that it would stand no chance against special 302-cu in (4.95-liter) Mustangs. So instead, Chevrolet combined the existing 327-cu in (5.36-liter) cylinder block with the 283 crankshaft, then added a large Holley four-barrel carburettor, L79 cylinder heads from the Corvette and an aluminium high-rise intake. Those heads had big 2.20-in (5.59-cm) intake valves and 1.6-in (4.06-cm) exhausts, with dual exhaust ports, and the ignition was boosted with high rpm distributor points. The result was a high- revving motor that produced 350hp (261kW) at 6,200rpm, and 320lb ft (433.9Nm) of torque. In the event, it was a more conservative 290hp (216kW) and 290lb ft (393.2Nm), but the effect was the same and it was enough to win races.

In theory, a buyer could order the Z28 engine on his new Camaro for an extra $358. In practice, the high-powered V8 came with a set of compulsory options such as heavy-duty suspension, close-ratio four-speed manual transmission, 3.73:1 rear axle, and power-assisted front disc brakes. Thus the complete car cost a hefty $4,100 at a time when the basic price of the hardtop Camaro was $2,608.

But that did not seem to matter. Trans Am racing stipulated that a

BELOW & OVERLEAF: The legendary Z28.

minimum number of cars be built, so that in theory Joe Public could go to his local dealer on Monday morning and order the same model that had won the weekend race. In practice, most of these cars (and certainly all of the early ones) went to established racers. They had to, to make it in time for the first race of the season. The first production Z28 rolled off the line on 29 December 1966. Roger Penske, the high-profile Chevrolet dealer and racer, took delivery of his on 10 January 1967 and immediately began to prepare it for the Daytona round of the Trans Am, then only ten days away. It was not quite a fairy-tale debut: Penske led for a while before retiring with engine problems, but Canadian Craig Fisher finished second in his Z28, ahead of two Mercury Cougars and a Mustang. Weeks later, it was announced that an SS Camaro convertible would be pace car for that year's Indianapolis 500, the dream exposure ticket for any muscle car market. By the end of its first year, just over 200,000 Camaros had been sold: not quite Mustang figures, but good enough. Really, the Camaro could hardly have got off to a better start.

For 1968 the Z28 got a new induction system with twin four-barrel carburettors, the intake carefully designed to send peak waves of fuel/air mixture into the cylinder at precisely the right time. Available as an add-on part for existing Camaros, it was in theory road-legal, but in practice its sole raison d'être was success in Trans Am racing. Chevrolet rated the new package, again conservatively, at 280bhp (209kW), though *Road & Track* considered that 350bhp (261kW) at 6,200rpm was a more realistic figure.

The magazine was probably right, as the Z28 proved marginally quicker than the equivalent Mustang, which was lighter in weight. The 0–60-mph time came up in 5.3 seconds when compared with the Mustang's figure of 5.4 seconds, and the standing quarter-mile in 13.77 seconds to the Mustang's 13.96: it also outpaced the

Ford by just 1mph (1.6km/h) on top speed, at 132mph (212.5km/h). With this much performance, the old criticism of axle hop was back, and the ride was described as stiff (not unexpected, in a race-bound car) and not everyone liked the gear-change. Still, everyone was impressed with the brakes (the Z28 needed 39 ft/12m less than the Mustang to panic stop from 80mph/129km/h) and power steering. The bottom line was that a Z28 won the championship that year.

While race-prepared Camaros tore up the tracks and drag strips, the design team began to plan variations for the road. An attractive Kammback, a sort of high-performance estate, got to the mock-up stage but was rejected as being too expensive to produce. Similarly, a fastback Camaro was rejected by the top brass, but this time because they had been tempted into building fastbacks before, but had failed to secure the anticipated sales. For the time being, the Camaro would have to soldier on as a hardtop or convertible only.

None of this was hurting its popularity. In 1969, the Camaro was voted top sporty car by readers of *Car and Driver* magazine. An advertisement of the time celebrated the fact with a to-the-point headline: "A word or two to the competition: You lose." Chevrolet could be forgiven for crowing over the Camaro's success. That year sales boomed to 235,147. Of those, over three-quarters were V8-powered, underlining the fact that straight-six "muscle cars," if they can be described as such, were often seen as price leaders. Their main job was to attract buyers into the showroom, whereupon the salesmen would do their job of talking customers into buying an optioned-up V8, which meant fatter profit margins all round. The same thing was true of the Z28, which made up only 3 percent of total sales: it too had to tempt potential Camaro buyers into their Chevrolet dealer, thanks to its exploits on the racetrack.

One thing that probably added to the Camaro's wide appeal was its styling, which was not overly aggressive, and the fact that women were often shown at the wheel in promotional advertisements; consequently, women were one in four Camaro customers. Actually, there were signs that Chevrolet thought it had gone a little too far in that direction, and the 1969 Camaro was given a little more aggression, with wider tires, bigger arches, and horizontal lights. Meanwhile, a three-speed Turbo-Hydramatic transmission replaced the old two-speed, and the manual option was four-speed. There was also the exciting ZL1 V8 engine, a 427-cu in (7.0-liter) all-aluminium unit that promised more power than ever before. But it was no Z28 replacement, being intended purely to challenge the Boss 429 Mustang in drag racing. Just 70 were made.

A new era for the Camaro began in February 1970, when Chevrolet general manager Pete Estes, who had overseen the Camaro launch, was replaced by the dynamic John DeLorean. DeLorean, of course, would much later become associated with a much less successful car, but in 1970 he was on the way up GM's corporate ladder. And he was adamant about one thing: under his leadership, the Camaro would have to knock the Mustang from its pedestal, and become the U.S.A.'s best-selling sports car. He also had in mind the fact that the Camaro was a popular first-time car; among the under-25s no less than 40 per cent of Camaro buyers were purchasing their first car. Brand loyalty being what it was, many of those people would go on to buy another Chevrolet, and another, so it made sense to give a big push to the first-car favorite.

Announced in February 1970, the Mk 2 Camaro was designed to look lower, wider, and more expensive than the car it replaced. Where the old Camaro was almost subtle in its performance pretensions, the latest one was clearly aspiring to

an Italian look, certainly upmarket. It
worked as well. There were other touches
like hidden windscreen wipers, bucket
seats, and a redesigned instrument panel.
The lowest-powered six was dropped,
and the three standard engines were now
a 155-hp (115.5-kW) 250-cu in (4.1-liter)
six, and 307-cu in (5.0-liter) and 350-cu
in (5.74-liter) V8s of 200 and 250hp (149
and 156.5kW) respectively. In SS form, a
300-hp (224-kW) version of the 350-cu in
was standard, with a 350-hp (261-kW)
396 optional. For the Z28 there was a
new 350-cu in (5.74-liter) unit rated at
406hp (303kW).

The press liked it. "The first of a new
generation of American GT cars," hailed
Car and Driver. *Motor Trend* recorded a
7-second 0–60-mph time for the new
Z28, with a quarter-mile time of 14.9
seconds. Oddly, *Sports Car Graphic*
could manage only 8.7 and 15.3 seconds

Another view of a Z28 from the side.

with the same model. Whatever the truth, it was clear that the Camaro was swiftly becoming a four-seat family alternative to the Corvette, with much the same performance. At first, sales went well, so well in fact that three-shift working was introduced to keep up with demand: in April 1970 the Camaro actually outsold all its rivals. But overall, sales were down, and the new car only sold 148,301 in its first year, whereas the Mustang managed 170,000.

In fact, the whole future of the performance car appeared to be in doubt during the early 1970s. The U.S.A. was leading the world in an ever stricter regime of safety and anti-pollution legislation. GM began a policy of reducing compression ratios to accept low-octane fuel, and with power now quoted net instead of gross, it made the 1971 Camaros seem weedy indeed. The engines were the 250-cu in (4.1-liter) six at 110hp (82kW), 307-cu in (5.0-liter) V8 at 140hp (104kW), 2bbl 350-cu in (5.7-liter) V8 at 165hp (123kW), 4bbl 350-cu in (5.7-liter) V8 at 210hp (156.5kW), SS 396-cu in (6.5-liter) V8 at 260hp (194kW) and 350-cu in Z28 at 275hp (205kW). There was a proposal for a new 400-cu in (6.55-liter) V8 to power the Z28 but this was turned down. The reality of the time was that car sales were falling and money was tight; in 1972 fewer than 117,000 Camaros found buyers.

In an attempt to drum up some interest, the marketing department put together three new packages, selling the Camaro as a four-seat sports car that could do everything a Corvette could, but for less money. Cheapest of the new set-ups was the Budget GT, which bought the 165-hp (123-kW) V8, four-speed transmission, F41 sports suspension, power steering, and a few other extras for $3,850. Then there was the Luxury GT, with all the interior touches like air conditioning, tinted glass, and a tilting steering wheel, for $4,365. Finally, the Performance GT was basically a Z28 with automatic transmission and Rally Sport interior, for $4,558.

It was not enough. The last straw was new standards for fenders, which looked like costing a great deal of money to apply to the Camaro. The car was selling reasonably well by pony car standards, but not in terms of other GM models. In mid-1972, Ed Cole made the announcement: the Camaro would have to die. But at the eleventh hour Chevrolet engineers found a way of meeting the fender standards at lower cost, by using extra struts and guards. After much argument, the Camaro was given a reprieve, and the threat of cancellation was lifted. It was just as well,

OPPOSITE:1969 Z28.

BELOW: The 1974 big-fendered Z28 Camaro.

since only 70,809 Camaros were sold in that year, though this was due in part to a long strike. Some half-completed cars were actually scrapped, as they had not sold and would not meet the 1973 safety and emissions standard.

The year 1973 was a slightly better one for the Camaro, with just over

90,000 sold, of which one in ten was a Z28, despite a power drop to 245hp (183kW), while the entry-level 250-cu in six was down to 100hp (74.5kW). But there was more equipment all round, with sports steering wheel, new seat harness, and floor shifter across the range. A new luxury Camaro, loaded with equipment, sought to tempt older, more affluent buyers: the Luxury Touring featured the 350-cu in V8, power steering, remote-control mirrors, special paint, trim, and instrumentation, plus 14 x 7 Rally wheels. It was little more than tinkering, but it seemed to work.

Nineteen-seventy-four arrived, when the Camaro seemed to be recovering from its sales low-point. First, DeLorean left Chevrolet, promoted even further up the GM greasy pole after his four-year

RIGHT & OVERLEAF: An early-1970s Camaro coupé. The wide wheels and tires are later editions.

stint, during which the marque had broken all sales records, overtaking Ford in the process. Then there was a whole new raft of safety and emissions standards: an ignition cut-out until seatbelts were fastened, for example, and in the following year the catalytic converter was introduced. In the Camaro, catalysts did come along with some other improvements, as well as a cleaner exhaust: electronic ignition was part of the package, which gave better economy and longer service intervals. The car was improved too, with power steering on all the V8s, a bigger fuel tank, and new air conditioning in 1974. The year 1975 saw a big wraparound rear window and many minor changes, while the RS was reintroduced: it had been dropped the year before, which had turned out to be a mistake.

But what of the 1974 oil crisis? Surely rocketing fuel prices and a 55-mph (88.5-km/h) speed limit really did spell the end of cars like the Camaro? Oddly enough, the opposite happened. By European standards, the Camaro is a big gas guzzler, but Americans saw it as a relative compact, which actually used less fuel than the average big sedan. As a result, sales soared to over 135,000 in 1974, and there was a repeat performance the following year. So high was demand that another GM factory, at Van Nuys in California, was turned over to Camaro production.

Times were changing fast, and Chevrolet advertising reflected the fact. "Camaro lets you limit your speed without cramping your style," went one advertisement. "Look good and feel good at 55…," which of course referred to the "double-nickel" speed limit, not the average age of Camaro buyers. "As long as you've got to go slower, you may as well do it in style." The whole language of car advertising had changed. The handling was no longer "race-bred" but "smooth and stable," and the Z28 was not "eye-popping" but "heavy-duty." It seemed like a complete culture

shift for Detroit: rubber-burning, gas-guzzling monsters were out, cool, safety-conscious cruisers were in. As if to underline the fact, the Z28 was dropped altogether in 1975.

It was a short-lived change of heart, however. In 1976 Chevrolet began to make advertising capital out of the International Race of Champions (IROC), in which well- known drivers from all types of motor sport raced against each other in identical Camaros, with the result that IROC became irrevocably linked with the car. Sales climbed to over 160,000 in that year (still not the happy hunting ground of the late 1960s, but a whole lot better than the low point), not far behind those of the Mustang. In the following year, there were actually power increases (5hp/3.7kW) across the range, but the big news was that the Z28 was back. "The Camaro Z28 is intended for

A 1978 Camaro Z28.

åthe macho enthusiast …a special breed of aspiration car…aggressive, quick, agile and dependable." It was clearly back to business as usual!

The new Z28 was not as quick as the old one (*Car and Driver* recorded 8.6 seconds to 60mph and 15.35 seconds for the quarter-mile) but the motoring press

There was a wrap around window for the 1975 Camaro.

was delighted to welcome back an old friend. Under the bonnet was the latest 350-cu in (5.74-liter) V8, bolted to a Borg-Warner four-speed manual transmission, with new suspension system incorporating stabilizer bars, different spring rates and rear shackles. Chevrolet sold over 14,000 examples of the new Z28 in that year, which helped boost sales again almost 200,000. Meanwhile, Ford sold 161,000 Mustangs, so the Camaro had finally fulfilled its original brief to outsell its arch-rival.

It was time for another facelift, and 1978 saw a freshened-up Camaro, with soft front end in body color, incorporating the bumper, and the same treatment at the back with wide tail lamps. Reflecting the times, the "Budget" had disappeared from the line-up, which now comprised the Sport Coupé, LT, RS, Rally Sport LT, and Z28. The last had a T-hatch roof, aluminium wheels, and a power boost to 185hp (138kW). *Car and Driver* now put it at 7 seconds dead for 0–60mph, but slightly slower over the standing quarter-mile at 16 seconds. The two millionth Camaro was made that year, and tuner Bill Mitchell was offering a turbocharged version. And it was the best sales year yet, with 247,437 cars finding homes, the Mustang trailing with 180,000. The good news for Camaro lovers continued in the following year, with a new luxury Berlinetta replacing the LT, while the EGR system was changed to improve economy and driveability. Sales were down slightly, to just over 230,000, but the Chevrolet M-car seemed to have put the 1974 oil crisis way behind it.

Then in 1980 came another oil crisis. There were few changes to the cars in that year, unless one counted new striping on the Z28, and sales slumped to barely 131,000, with another bad year in 1981. But the old Camaro was only treading water, because a new one was on the way.

A Camaro for the 1980s

When the new Camaro was being planned in the troubled 1970s, one thing was crystal clear: to survive, it would have to be smaller, lighter, and more fuel-efficient than the old one. But it also had to retain a family resemblance, as Camaros were now part of the fabric of American motoring society, driven by an entire generation; one does not reject that sort of heritage lightly.

When it was finally unveiled in 1982, the new car did represent a radical change, but not as radical as it might have been. All sorts of things had been considered in the quest for lightweight efficiency: front-wheel drive, sliding doors with fixed glass, short, squat front ends. As it turned out, the new Camaro had none of those things, but the new wedge shape clearly contained Camaro genes. It looked thoroughly modern, but also slightly retro, the big glass hatchback styled like a traditional notchback coupé. The big rear glass, with its double bend, was a challenge in itself, and the windshield was raked back at 63 degrees, in keeping with the car's overall wedge shape.

Under the bonnet, the base engine was a 2.5-liter four-cylinder unit – a mere four cylinders in a Camaro! It was enough to make the 1960s gas heads spinning in their graves. Of course, it was not any old four, having the benefit of throttle body fuel injection, and if that were not enough, next up came a 434-cu in (2.8-liter) carburettor-fed V6, and if one really insisted, there was a 775-cu in (5.0-liter) 4bbl V8 in normally- aspirated and fuel-injected versions. All came in that cheese-wedge body, with body-colored soft-nosed fenders of a urethane-covered honeycomb. Despite being smaller, the new car was roomier than the one it replaced, though there was less luggage room, and had better visibility. Had the Camaro gone all sensible?

There were just three models now, namely the Sport Coupé, Berlinetta and Z28, all of which gained a five-speed manual gearbox and four-speed auto in 1983, an option on the SC, standard on the other two, while the ultimate 5.0-liter V8 on the Z28 now had Cross-Fire fuel injection. That year, *Motor Trend* voted it Car of the Year, and the Camaro continued to lead its class in sales. As for the Z28, purists may have looked down on it as a lesser descendent of the original, but in reality it was just as fast: with the L69 High Output (HO) V8 it produced 190bhp (141.5kW) at 5,800rpm, which allowed the 0–60-mph dash in 7.2 seconds. It could accelerate from 0–100mph (161km/h) and brake back to a stop again in 23.8 seconds. The HO was the option to have for performance freaks, with high-lift camshaft and improved ignition, intake, and exhaust systems.

A decade after that first oil crisis, performance was back with a vengeance, and for 1985 Chevrolet announced the IROC-Z28, based on the cars supplied for the IROC series, which apart from a break between 1981 and '83, was still ongoing and still using Camaros. Lowered all round, the IROC-Z28 used gas-filled Delco/Bilstein shock absorbers and ventilated disc brakes. The fuel-injected LB9 V8 was not more hugely powerful than the standard Z28, but with 215bhp (160kW) and 275lb ft (372.9Nm), it was fast enough: *Motor Trend* reckoned on 6.87 seconds to 60mph and 15.32 for the standing quarter-mile. That performance was complemented by a high-ratio power steering set-up (2.5 turns lock to lock) and 245/50-16 tires. Both Sport Coupé and Berlinetta continued with the V6, now with fuel injection.

Performance was back in fashion, and the IROC-Z28 proved most popular, doubling its sales in 1986. Meanwhile, the base Sport Coupé was powered-up with the 171-cu in (2.8-liter) V6 with multi-port injection, wide wheels, and five-speed gearbox. It was also the last year for the luxury Berlinetta, suggesting that the

Camaro might be returning to its performance roots: in fact it was, for nearly half of all Camaro sales now consisted of the Z28 and IROC-Z28.

Sensing that horsepower was back in fashion, Chevrolet uprated the IROC-Z28 in 1987, with the L98, a 348-cu in (5.7-liter) fuel-injected V8 straight out of the Corvette. Now with 220hp (164kW) or, according to some, 225hp (168kW) at 4,200rpm and 320lb ft (433.9Nm), the fastest Camaro could reach 60 mph from rest in just 6.3 seconds, and took 14.5 seconds for the standing quarter-mile. Apart from a 3.27:1 Australian-built Borg-Warner rear axle, the running gear was the same

OPPOSITE: A 1978 model. A T-roof replaced the full convertible.

ABOVE: A 1984 Camaro.

as in the standard 5.0-litre version. But just as with the original Z28, one could not buy the engine option alone. It came with a whole list of mandatory "options:" upgraded four-speed auto transmission, limited-slip differential, four-wheel disc brakes, and engine oil cooler. Together, these almost doubled the price of the

ABOVE: A 1985 model.

OPPOSITE ABOVE: 1988 Coupé.

OPPOSITE BELOW: A convertible Camaro returned in the late-1980s.

OVERLEAF: A Camaro Z28 for 1986.

engine, but *Hot Rod* magazine thought the result was worth it: "It could be," a journalist wrote, "the closest facsimile to a full-bore road-racing car you'll ever drive."

The Camaro legend is now fifty years in the making. The new 2017 range has been refined for today's generation.

PONTIAC FIREBIRD/TRANS AM

In many ways, the Pontiac Firebird is the forgotten muscle car. Like every other non-Ford pony car, it was a reaction to the Mustang. General Motors' pony car was really a Chevrolet project, which Pontiac tagged on to late in the day; for the first few years of its life it played second fiddle to its GM sibling, the Chevrolet Camaro. The Firebird, moreover, did not sell as well as its Camaro cousin, let alone the Mustang itself.

And yet the Firebird turned out to be the great survivor among muscle cars, persevering with a big, hairy 455-cu in (7.47-liter) V8 in the uncertain 1970s, when every other muscle car was downsizing, downgrading, or hiding its horsepower beneath a bushel. It was all due to the Trans Am, which started off as a

Firebird derivative but ended up as a model in its own right. There was a paradox here too, for the Trans Am, one of America's legendary hot cars, was named after the famous race series but was never a successful racer itself. Yet up to 2002 one could still buy a brand-new Firebird or Trans Am.

In 1964 the Mustang had been let out of its stable and, contrary to the predictions of the industry, was selling in huge numbers. Over at General Motors, Pontiac top management was peopled by individuals who loved and understood performance cars, most notably Bunkie Knudsen and John DeLorean. In a few short years they had helped to change Pontiac's image from fuddy-duddy to hot and desirable, with the Super Duty racing specials. On the back of that, they had recently launched the GTO, which was destined to become the first of a new generation of muscle cars. If any GM division was equipped to meet and beat the Mustang, it was the newly performance-aware Pontiac.

They got to work. Chevrolet had already proposed the XP-836, a sporting four-seater, but John DeLorean in particular wanted a proper two-seat sports car,

OPPOSITE & RIGHT:
The Firebird was introduced
in 1967 by Pontiac. This
coincided with the release of
Chevrolet's Camaro.

a cheaper competitor for Chevrolet's own Corvette. This was the XP-833, with a streamlined, futuristic plastic body. But it was too avant-garde for GM's top management, who in any case did not want any in-house competition for the Corvette. Consequently, the XP-833 died a swift death.

Meanwhile Chevrolet has been working hard on the XP-836, the car that would become the Camaro. GM decreed that instead of producing its own Mustang rival, it would work with Chevrolet on the XP-836, which subsequently became a joint project. This put Pontiac at a disadvantage, as Chevrolet was already some months down the design road, and coming to the project so late meant that Pontiac would have little influence on the car's fundamentals. "Pontiac didn't like the decision," wrote Bill Holder and Phillip Kunz in their book *Firebird & Trans Am*, "but that was the way the game would be played." The Camaro's styling had already been finalized, and to cut costs (and time) the new Pontiac would have to share its wings and doors: only the nose and tail could be altered to make the car different in character from the Camaro.

They managed to do this by giving it a GTO-style split-front grille with recessed twin headlights and narrow rear lights in two tiers, instead of the Camaro's conventional lights. It was not much, but at least the cars now looked like cousins rather than clones. The downside was that the extra design work pushed the Pontiac's launch date back to February 1967, five months after that of the Camaro. However, the Firebird did have a suitably evocative name.

Late Starter

"After this," went the advertising copy, "you'll never go back to driving whatever your driving." That ran below an almost full-page color picture of an open-top Firebird 400 roaring along an open mountain road, devoid of traffic (in car advertisements, roads never have any other traffic). The advertisement also referred to "400 cubes of chromed V8...heavy-duty 3-speed floor shift, extra-sticky suspension, and a set of duals that announce your coming like the brass section of the New York Philharmonic." If Pontiac was attempting to sell a politically incorrect fantasy, it was doing a pretty good job.

That 400 was the top model of five: Firebird, Sprint, 326, 326HO, and 400. The base model came in at only $2,600, powered by Pontiac's own overhead-cam 230-cu in (3.77-liter) straight six with 165hp (123kW). Despite all the components it shared with the Camaro, the Firebird used Pontiac's own power units. That also allowed Pontiac to offer the Sprint, which was surely unique among muscle cars in using a highly-tuned six-cylinder engine instead of a V8. For only $116 extra over the base Firebird, the straight-six was given a Rochester four-barrel carburettor, higher-lift cam, 10.5:1 compression ratio, split exhaust manifold, and freer-flowing air cleaner, plus 215hp (160kW), 50hp (37kW) more than standard. As part of the package, the three-speed shifter was floor-mounted, and the suspension was firmed up. By American standards, this was a relatively small-engined high-revving performance car. Both the Sprint and base Firebird could be had in open-top forms, as a $237 option. Despite that seductive advertisement, convertibles were only ever a minority of Firebird sales: less than 16,000 were sold in that first year, or about one in five of the total.

If the Sprint seemed too frenetic, and for many traditionalists it probably did, the more laid-back 326-cu in (5.34-liter) V8 was a better choice. Slightly more powerful than the Sprint at 250hp (186.5kW), but with far more torque as it provided 330lb ft (447.5Nm) at 2,800rpm, this was the most relaxed of the new Firebirds. Pontiac underlined the point by equipping it with a non-sporting three-speed column shifter, and billing the buyer for $21 less than the Sprint owner. A three-speed transmission was not synonymous with boulevard cruising: even the top-performance Firebird 400

ABOVE: The distinctive Firebird Badge.

LEFT: 1969 Pontiac Firebird was Pontiac's version of the Camaro. But it came to the project late, and could only make limited changes to Chevrolet's design.

stuck with a three-speed. For an extra $47, Firebird 326 owners could make that a floor shift, while a center console and bucket seats were also on the options list. As ever, the interior design of mass-market muscle cars was just as important as the way they looked on the outside. Owners were paying for something that made them feel special. Not for nothing had the Camaro's designers aimed for a fighter plane feel for the interior.

But it was the two top Firebirds, the 400 and the 326HO, that finally entered true muscle car territory. The HO took the standard 326-cu in engine and added a 10.5:1 compression, Carter four-barrel carburettor, and dual exhaust, among other things. According to Pontiac, this boosted power by a modest 14 per cent to 285hp (212.5kW). This figure is generally considered to have been underrated, with the true figure at 300hp (224kW) or more. The official torque figure seems more believable, at 359lb ft (486.8Nm). To proclaim to the world that one had bought an HO, the car sported long body stripes and "HO" badges: in this instance, the HO stood for High Output.

At the end of that first 1967 model year, 82,560 Firebirds had found buyers, which was surely not as many as Pontiac would have liked, but the car was hamstrung by missing the first five months of the sales season. That was partly why Chevrolet managed to sell over 200,000 Camaros in the same year. Both were a long way off Mustang figures, but the figures were good enough. And in 1968, the Firebird's first full year on sale, 107,000 cars were sold. In fact, 1968 would head the Firebird's sales record for eight years. This, of course, was the height of the muscle car boom; only a few short years into the future and Pontiac salesmen would be looking back to these times with feelings of nostalgia.

There were few major changes in that year. The straight six got a capacity boost to 250 cu in (4.1 liters) which gave 175hp (130.5kW) and 240lb ft (325.4Nm), though it was still a mildly tuned unit, with a lowly 9.0:1 compression ratio. Oddly, although the Sprint enjoyed the same increase in cubes, it was still quoted at 215hp (160kW), though rated torque was up to 255lb ft (345.8Nm): maybe the original had proved just a little too "European" for traditional U.S. buyers. Of more significance was the replacement of the 326-cu in unit with the Firebird 350. Like its predecessor, this 350-cu in (5.74-liter) V8 came in two forms: mild-mannered cruiser with a single two-barrel carburettor, three-speed column shifter and now 265hp (197.5kW) at 4,600rpm, and as the 350HO with four-barrel Rochester, 10.5:1 compression, and 320hp (238.5kW). As with the original HO, there were plenty of cues on the outside as to what lay beneath the bonnet, such as dual exhaust and F70 x 14 tires, plus all the usual stripes and badges. The 350HO's power output was not far behind that of the 400, now boosted slightly to 335hp (250kW) at 5,000rpm as the 400HO and still offered with or without Ram Air. This was due to a higher 10.75:1 compression and power-flex fan. Any performance difference between the 350 and 400 engines may have been small, but for many buyers there was still one very good reason to pay the extra $435 for a 400: it still had more cubes than any other pony car – even the latest 1968 Camaro had only 396. Of course, that could be coaxed up to 375hp (279.5kW) with a dealer-fitted hot camshaft. The Z28 Camaro produced 400hp (298kW), but for many buyers, the Firebird 400 remained the cubic-capacity king. To maintain interest, Pontiac offered an updated Ram Air system, the Ram Air II, late in the model year, though the real significance lay not in the Ram Air itself, but in the fact that the new engine was significantly stronger than the old one, with forged pistons and four-bolt main bearings.

Any doubts Pontiac may have had about adopting and adapting what was basically a Chevrolet design would have been banished by a six-car test in *Car and Driver* during May 1968. A Firebird 400HO was compared with a Camaro SS396,

AMC Javelin SST, Mustang 2+2 GT, Mercury Cougar XR-7, and Plymouth Barracuda Formula S. The Pontiac product came out head and shoulders above them all.

Car and Driver loved its big engine, though admittedly it had been very well prepared, and scored it top of the six in every category. It was the fastest-accelerating of the six, and by some margin, attaining 60mph (96.5km/h) in 5.5 seconds when compared with the 6.6 and 6.3 seconds recorded by the Camaro and Mustang respectively, and the quarter-mile time of 14.2 seconds. Moreover, it

OPPOSITE: Two-piece rear lights distinguished the Firebird from the Camaro.

revved so quickly and smoothly that it was easy to overshoot the 5,100rpm shift point. The variable-ratio power steering (the first on an American-made pony car) came in for particular praise, as did the handling. There was no mention of the axle hop which early Firebird tests criticized, so the changes for 1968 (multileaf springs and staggered shocks) seem to have worked. The ride was thought to be a little too firm, but the Firebird scored bottom only on front fender protection and the effectiveness of its wipers, hardly major points. When all the points were added up, the Firebird scored 118, ahead of the Barracuda (111), Javelin (91), Cougar (90), Camaro (79) and Mustang (73).

"For sheer enjoyment and confidence behind the wheel," the *Car and Driver* testers concluded, "the Firebird was almost in a class of itself." By contrast, the Camaro was "built to be all things to all people, and as a result, it was a disappointment." That surely was the sweetest victory for the Pontiac engineers. They had succeeded in taking a Chevrolet design and transforming it through careful use of their own components. As *Car and Driver* acknowledged, the division seemed to have a knack for taking unpromising material and producing something very different. If *Car and Driver*'s test was to be believed, it certainly succeeded with the Firebird.

The Arrival of the Trans Am

It had not escaped the Pontiac management that the Mustang had come last in that *Car and Driver* test. The reason was simple: the Mustang had barely changed in three years, while newer rivals were coming thick and fast. They were determined that no such fate should befall the Firebird, so for 1969, after less than two years on the market, the Firebird received some major changes. As well as the subtle alterations to the sheet metal enjoyed by the Camaro that year, the Firebird's front end was restyled, with the quad headlights now carried in body-colored moldings, with the famous split grille squeezed inward to make room. Inside, there was an improved dashboard and new safety features.

The Sprint got a power boost to 230hp (171.5kW), but the best-seller of the entire range remained the softer Firebird 350, its power unchanged. The 350HO could boast an extra 5hp (3.75kW), which was not much but which counted on the all-important specification sheet, to 325hp (242kW), as a result of new cylinder heads, larger valves, and a higher lift cam. The base 400 now offered only 5hp more than the 350, but 13 per cent more torque at 430lb ft (583.1Nm). It was also given all the visual cues of the largest-engined Firebird on offer: hood scoops (which was available even without Ram Air), dual exhausts, and floor shifter. An extra $435 brought Ram Air III, which gave those hood scoops something to do, and according to Pontiac delivered an extra 5hp and identical torque to the standard 400. If that was not enough, there was a new Ram Air IV for 1969. It was expensive at $832 extra (and not many were sold) but promised much, with a hotter cam, aluminium cylinder heads, and different valve train. Another clue as to why not many Firebird customers chose to tick the Ram Air IV box lay in Pontiac's official power figures, which put Ram Air IV at 345hp (257kW) at 5,400rpm, with identical torque. Most considered that 10hp (7.45kW) more for $800 was not an outstanding deal. However, the very same engine in Pontiac's GTO was quoted at 370hp (276kW). Either Pontiac had made a slip, which was unlikely, was being modest, which was very unlikely, or wanted to take the heat off its top-performance car in the face of the criticisms of safety campaigners.

Whatever the truth, it was overshadowed by a new option for the Firebird that, with hindsight, was more significant than any of those: the Trans Am. Mention that name today, and most people will not think of the race series for pony cars, but a hot Pontiac that made the name its own. Pontiac's plan went like this: it did not have a suitable competitive engine for the popular Trans Am series, so it started to develop a 303-cu in (4.965-liter) V8 based on the 400. In the meantime, Firebirds were raced using Z28 Camaro power units. After racing a while, the special 303 would be fitted to a road-going Trans Am named after the series. But it took too long to develop, and by the time 303 was ready, the rules had changed, and it was not legal for Trans Am racing.

Faced with race-heritage rivals like the Z28 and Boss Mustangs, moreover, Pontiac did not want to delay its road-going spin-off, so the Firebird Trans Am was launched anyway, at the Chicago Auto Show in March 1969, though powered by the existing 400-cu in motor rather than the special 303. Consequently, when it was launched, the new Trans Am had never actually turned a wheel on a race track. Pontiac even had to pay the Sports Car Club of America a $5 royalty on every car. Still, it was a good investment, given that the car was still going strong 30 years later and was better known than the race series!

So what was the basis of the Trans Am? It seems almost sacrilegious to suggest it, but without that special 303-cu in race-bred engine, the early Trans Am was little more than a Firebird 400 with a spoiler and a different paint job. Engine-wise it was identical to the top 400s, with 335-hp (250-kW) Ram Air III or so-called 345-hp

BELOW & OPPOSITE: A 1973 Trans Am with a SD-455 engine.

(257-kW) Ram Air IV. It looked very different, however, and this was part of its reason for being. John DeLorean was the power behind the Trans Am. As GM's Pontiac chief, he could not help but notice how well the Z28 Camaro had been selling. What the Firebird needed was its very own Z28, and the project that was initiated was codenamed "Pontiac Firebird Sprint Turismo."

When launched, the Trans Am looked as though it would be loud even before it was started up. All the early cars came in white, with blue racing stripes running the length of the hood, roof, and trunk lid. Two gaping scoops gaped hungrily out of the hood, two more extractor scoops were fitted just behind each front wheel and there was a large rear spoiler, which the engineers calculated to produce 100lb (45kg) of down-force at 100mph (161km/h). So the Trans Am was flashy (tame by the standards of the 1970s, but a rabid extrovert for its time), but there were some suspension changes over the standard Firebird as well, namely heavy-duty springs, a 1-in (2.54-cm) stabilizer bar, and 7-in (17.8-cm) wheels. It was not a mere paint and badges job, but it was close.

Nor was the first Trans Am a mass-market car. Pontiac made only 697 in its first year, most of which were fitted with Ram Air III, and most were automatics. But despite the extra glamour of the Trans Am, Ram Air IV, and the restyling, Firebird sales as a whole were actually down in 1969, to 87,000, and that was over an extended 15-month model year, as a strike delayed the launch of the 1970 cars.

For 1970, the Firebird, and of course the Trans Am, shared new single headlight styling with the Camaro: the Chevrolet and Pontiac vehicles would share bodies until 1981. Most people liked the cleaner look, with more subtle hood scoops and an inbuilt rear spoiler on the top Formula 400. But the times were reflected in other changes. To rationalize and save money, Pontiac's overhead-cam six was dropped in favor of the equivalent Chevrolet six, which brought the power down to 155hp (115.5kW). The peppy Sprint was replaced by the Esprit, which made luxury, rather than performance, its top selling point. It was powered by the 350 V8, now with lower 8.8:1 compression and single two-barrel carburettor, producing 255hp (190kW). Meanwhile, the three 400-powered Firebirds were replaced by the Formula 400, offered with either 330-hp (246-kW) base-model engine or the 335-hp (250-kW) Ram Air III. Ram Air IV was now only available to special order. The Formula's suspension was upgraded along Trans Am lines, with front and rear stabilizer bars and heavy-duty springs.

The 1970 Trans Am ("1970+" in official Pontiac jargon as it had been launched late, in March 1970) went along with the more subtle approach of the Firebirds. The rear spoiler was the muted, inbuilt unit seen on the Formula, and new front spoiler (giving up to 50lb/23kg of downforce according to Pontiac) was also smoothly styled in. The standard power unit was Ram Air III; here again Ram IV was available on special order, but was taken up by only 88 Trans Am buyers. Nevertheless, they still got a Shaker hood (in place of the twin front-facing scoops), Rally II wheels, power brakes, and steering, concealed wipers, and dual horns, while the front stabilizer bar was a beefier 1.25in (3.175cm). The changes worked, and more than 3,000 Trans Ams were sold in that year. This was not enough to worry Ford or anyone else, but was a useful boost to the Firebird, whose sales collapsed disastrously in 1970 to less than 46,000, or half the 1969 figure. The muscle car bubble had burst, but could the Pontiacs cope?

The 455

In many ways, Pontiac made the same efforts at retrenchment as every other muscle car manufacturer in the early 1970s. Compression ratios were scaled back to cope with lower octane fuel and stricter emission limits, and there was as much emphasis on luxury muscle (in the new Esprit Firebird, for example) as on sheer horsepower. The whole atmosphere had changed as horsepower figures tumbled. Much of the latter was the result of a change in measurement from gross to net horsepower, but it served only to reinforce the idea that the performance car was retreating.

In the midst of all this, Pontiac launched the biggest ever V8 fitted to a pony car, a 455-cu in (7.46-liter) unit. Fitted only to the Formula Firebird or the Trans Am, it came in two guises: 8.2:1 compression for 325hp (242kW) gross or 255hp (190kW) net, or 8.4:1 and a four-barrel carburettor HO with 335hp (265kW). Naturally, these massive motors pushed out a great deal of torque, and even the low- compression version managed 455lb ft (617Nm): in other words, one pound per foot per cubic inch, more likely to have been the result of careful development rather than sheer coincidence.

If the 455s seemed over the top, one could still buy a Formula 400 (now detuned to 300hp/224kW gross or 250hp/186.5kW net) with a lower-powered, budget-priced Formula 350 which cost only $29 more than the Esprit. The idea of the new, cheaper Formula was to provide all the performance add-ons (bucket seats, suspension package, hood scoops) with lower purchase, insurance, and running costs. With this car and the 455s, Pontiac was offering Firebirds in tune with the incoming tide of

safety and emissions legislation, but at the same time struggling to come to terms with the inevitable.

The Esprit, meanwhile, came with 350 or 400 two-barrel V8s, both in modest states of tune, and buyers of the basic Firebird had a choice of the 250-cu in (4.18-liter) straight-six or the 350. Not surprisingly, the Trans Am came only with the new 455 motor, but delivered 335hp at 3,500rpm and 480lb ft (650.9Nm). In fact, despite the hard times for muscle cars, there were more Firebird models to choose from than ever before. Not that it helped sales much. They were up marginally, to 53,000 (excluding just over 2,100 Trans Ams) but, to put that in perspective, the Pontiac was outsold by Camaro and Mustang by three to one.

The year 1972 proved to be the Firebird/Trans Am's low point: this was in fact the year that GM considered dropping both the Chevrolet and Pontiac pony cars, though both just managed to

1973 Trans Am SD-455.

survive. In keeping with the times, the best-selling model was the basic Firebird, which made up 40 per cent of the 1972 total. It still came with a choice of detuned 250-cu in six (now derated to 110hp/82kW net) or two-barrel 350-cu in (5.74-liter) V8 offering 160hp (119kW). It was closely followed by the Esprit (still majoring on chrome and interior fittings), now available with a 400 V8 as well as the 350. And the performance Formula? This had greater engine choice than ever before: 160- or 175-hp (119- or 130.5-kW), 250-hp (186.5-kW) 400 or 300-hp (224-kW) 455. Despite these options, only a little over 5,000 vehicles were sold, so despite the extra choice and the reflected glamour of the Trans Am, it seemed as though buyers really were abandoning muscle cars. The Trans Am itself came only with that top 455 motor, and about two-thirds of buyers opted for the M-40 automatic rather than the Hurst shifter four-speed manual.

The power figures seemed low, but all were net horsepower, so for performance freaks things were not as bad as they appeared: a Formula 455, for instance, could still run the quarter-mile in 14 seconds, so it was no slower than its predecessors. All well and good, but less than 1,300 Trans Ams trickled off the production line (a result in part of labor problems) and sales of the whole range amounted to 28,700. It was little wonder that the GM top brass considered dropping the whole lot.

But having decided to hang on in, Pontiac took no half-measures. For 1973, while almost every other muscle car was in retreat, the Firebird and Trans Am were given extra power. It came in the form of the SD-455 V8, which was the existing 455 with many tweaks and changes to produce 310hp (231kW) net. That was equivalent to around 350hp (261kW) gross, so this 1973 V8 was offering the same sort of power as muscle cars of the late 1960s. To underline a return to what some would see as the good old days, Pontiac resurrected the "SD" or "Super Duty" tag, the name given to all those drag racing special parts of the early 1960s.

SD was no misnomer, as it was a very different motor from the standard 455, with a reinforced block, forged rods, aluminium pistons, Quadra Jet carburettor, dry sump lubrication, special cam, four-bolt main bearings, and dual exhaust. In fact, so hot was the SD that it failed to meet emission limits, and Pontiac had to hurriedly derate it by 20hp (14.9kW). With a milder cam and cleaner exhaust it could manage 290hp (216kW), still more than any other muscle car. Moreover, the Firebird/Trans Am still had more cubic inches than any rival, and for some people these things still counted. In the Trans Am, the SD-455 came with beefed-up three-speed Hydra-Matic (with the shift point raised to 5,400rpm), a heavy duty radiator, and uprated suspension. Oddly enough, having produced a hot car in a cold era, Pontiac seemed a little shy of the fact. One Trans Am 455 buyer recalled that he "had to beg" for his car, and wait five months for delivery. And he also had to pay $675 extra for the privilege, which was quite a chunk of money in 1973. Pontiac's reticence was at odds with the hottest Trans Am yet, especially with its new for 1973 big bird decal, which now covered the entire hood.

Originally, the SD-455 was to have been a Trans Am option only, but Pontiac decided to offer it on the Firebird Formula as well, which continued with the standard 455, 400 and 350 V8s. As ever, the Formula offered a good-value performance package: for only $27 more than the Esprit, buyers got a twin-scoop hood, heavy-duty suspension, dual exhausts, and several other bits and pieces. But as if to atone for the SD, other Firebird powerplants were derated in that year, with lower compressions: the base 250-cu in six was down to 100hp (74.5kW), and only 10 percent of base Firebird customers chose it. Most went for the torquier 350 V8. In fact, the best-seller in 1973 was not that one, nor the sporty Formula, still less the flamboyant Trans Am, but the luxury Esprit. All told, Pontiac produced just over 46,000 Firebirds and Trans Ams for 1973, so the SD had not reversed the decline.

In fact, 1974 would be the SD's second and final year. Although applauded by the performance freaks, it really was swimming against the tide: gas was getting expensive, and it almost seemed unpatriotic to waste the stuff. Less than 1,000 Trans Am SDs were sold in 1974, and a mere 58 Formula SDs (evidently there were not enough performance freaks around). But sales of less macho Firebird/Trans Ams actually rose that year, to well over 70,000. Once again, the luxury Esprit was the best-seller, followed by the base Firebird, the Formula and, bringing up the rear, the Trans Am. The Trans Am package had become comprehensive, and was no rebadged Firebird: power steering, front discs, limited-slip differential, Rally gauges, Rally II wheels, sport suspension, and many other parts were all standard.

RIGHT & PAGE 166: Other muscle cars fell by the wayside, but the Trans Am struggled on.

Even with the milder-sloping front end fitted to all Firebird/Trans Ams in that year, it looked the part.

There are two ways of looking at the Firebird/Trans Am in the mid-1970s. Enthusiasts would bemoan the dropping of first the SD-455, then the basic 455 (though it would return in derated form), with ever lower compression ratios and power outputs for the other engines. On the other hand, Pontiac were selling more Firebird/Trans Ams than ever before. In 1974, the Trans Am became the best-seller, with over 27,000 sold, while the following year combined sales exceeded 110,000, a significant milestone. There were few important changes to the actual cars, more a juggling of engine options to give more choice to base and Esprit buyers, and more emphasis on the Formula's appearance than its quarter-mile times. The base Firebird, for example, actually got a slight power boost in 1976, to 110hp (82kW), the choice of 160- or 175-hp (119- or 130.5-kW) 350 V8s and, for the first time, the 185-hp (138-kW) 400. Meanwhile, the Trans Am was still the best-selling model, with over 46,000 finding homes. Even more significant for Pontiac's sibling rivalry with Chevrolet, the Trans Am was now outselling the Corvette. In short, it had become a muscle car model in its own right, and from now on, "Trans Am" would be just as recognizable as "Firebird."

Boom and Bust

In the late 1970s both the Trans Am and Firebird enjoyed a sales boom, with over 150,000 cars (almost 70,000 of them Trans Ams) sold in 1977 alone. Well over 200,000 found homes in 1979. It is hard to say why this happened. With the 455 V8 finally gone, the Pontiac pony was following all the other muscle cars down the road of lower compressions and fewer cubic inches, though looks must have had a lot to do with the car's success. Pontiac's pony still looked the part of a true muscle car, especially in Trans Am form or as the Formula with its optional W50 body package. Trans Am special editions in black and gold, or Firebird Esprits in bright blue or red, all with loud decals and stripes, were also popular.

Recognized by its four rectangular headlamps, Trans Am/Firebirds of this era also saw some significant engine changes. The faithful straight six was replaced by a 231-cu in (3.785-liter) V6, sourced from Buick, in two- or four-barrel forms. Meanwhile, a small-block V8 of 305 cu in (5.0 liters) was brought in from Chevrolet, offering 145hp (108kW). The familiar 400 continued, but was topped by a 403-cu in (6.6-liter) V8 with 220hp (164kW): it was not a 455, but certainly on the way there.

To celebrate its tenth birthday in 1979, the Trans Am enjoyed its best sales figure yet, at 117,000. On the tenth anniversary special edition, there was a bigger bird on the bonnet, silver leather bucket seats, and a silver-tinted hatch roof. The pure convertible had long since gone, replaced by an optional T-roof. The special edition also celebrated the fact that, once again, the Trans Am had been chosen as the official pace car for the Daytona 500; this always impressed the buying public, and Pontiac made the most of the fact with pace car replicas.

In the following year, however, both Trans Am and Firebird sales slumped. Was it because the big 400 and 403 V8s had been dropped? Possibly it was, though the 301-cu in (4.93-liter) turbo V8 which replaced them was almost as powerful, at 210hp (156.5kW). That 301 had become the standard powerplant, available in 140- and 155-hp (104- and 115.5-kW) forms alongside a single V6, and of course, the 210-hp turbo. Once again, there was a Daytona pace car replica, and over 5,000 of these were sold. But maybe the Trans Am/Firebird was simply looking old: after all, the basic design stretched back over a decade, and in 1981 sales had slumped again. Pontiac needed a new one.

It arrived in the following year, with a sleeker, cleaner all-new body shape that nevertheless managed to retain a family resemblance to the old Firebird. With more attention given to aerodynamics, there was a steeply raked windscreen and squared-

off tail, while the new car weighed 2,800lb (1270kg), significantly lighter than the old one. Pontiac was following Ford's example with the new Mustang: shaving off weight to keep up performance with less power and better economy than the old-style muscle cars. Like the Mustang, the base power unit was a four-cylinder 153-cu in (2.5-liter) motor, the "Iron Duke" with a mid-range 171-cu in (2.8-liter) V6 and, for the Trans Am, a 150-hp (112-kW) 305-cu in (5.0-liter) V8. For 1983, a V6 HO was added, with 135hp (100.5kW), while the V8 came with a choice of carburettor or fuel injection. There was a special edition too, the 25th Anniversary Daytona 500 Pace Car, plus the black-and-gold Recaro Trans Am with special interior and an extra $3,610 on the dealer's bill.

By and large, buyers liked the new cars, though sales did fluctuate through the 1980s. They soared the first year, dipped in 1983 then soared again to nearly 130,000. In 1985 they were down again, falling to less than 100,000.

There were just three models: base Firebird, S/E, and Trans Am, the last of which remained the best-seller, making up nearly half of all sales. The two top models were facelifted for 1985, with new front and rear ends, with the new look especially dramatic on the Trans Am with its optional W62 ground effects package. This was clearly inspired by racing, with aerodynamic "skirts" as well as the usual front and rear spoilers. They might have reduced ground clearance, but who cared? Perhaps more significant was the rebirth of performance.

Along with other muscle cars of the mid- and late-1980s, the Trans Am began once again to offer big horsepower figures. The 305-cu in (5.0-liter) V8 now came in three versions: base, HO, and TPI (Tuned Port Injection) with 155hp (115.5kW), 190hp (141.5kW) and 205hp (153kW) respectively. Performance, it seemed, was back in fashion. Even cubic inches were making a comeback, and for 1987 Pontiac fitted a 350-cu in (5.735-liter) V8 with 210hp (156.5kW). The engine was from Chevrolet, but Pontiac needed it to counter the new 5.0-liter V8 Mustang. It was hardly surprising to learn that the 2.5-liter four-cylinder Firebird had been dropped.

There was a multiplicity of options to go with the new-found performance. The Y99 suspension package, for example (front and rear stabilizer bars with custom shocks), and a whole range of rear axle ratios. For 1988, the 350-cu in TPI V8 was up to 235hp

(175kW), available both in Trans Am and a reborn Formula, while there was a new GTA, which slotted in midway between those two. This was a sort of performance Firebird, with a whole range of options like all-wheel disc brakes and performance suspension to turn it into a budget Trans Am. It was a popular car, with over 20,000 built. Once again, the Trans Am enjoyed pace car status in 1989, though the replica, of which more than 5,700 were sold, used a turbo V6 rather than one of the big-cube V8s. In actual fact, it was more powerful than any of them, with 245hp (183kW).

On the face of it, the Trans Am and Firebird seemed to have moved with the times. Lighter and more efficient than the old-school muscle cars, but now with big performance once again, from a choice of V8s offering 200hp (149kW) or more. They looked the part too, muscular, but aerodynamic, and clean. And yet the honeymoon was over. From 1987, sales fell year on year, plumbing even greater depths than the dark days of the early 1970s. Pontiac sold over 110,000 pony cars in 1986, little more than half that in 1989 and a little over 20,000 in 1990. Nor was the Trans Am the tower of strength it once had been: it became a minority seller in the range, with less than 2,500 sold in 1990. Sales did recover slightly in 1991, thanks to a complete restyle front and rear, and a new convertible option across the range, while the 5.7-liter TPI V8 came in 205-, 230- and 240-hp (153-, 171.5- and 179-kW) forms. For economy-minded buyers, there was still a 3.1-liter injected V6. But it was a temporary blip, and sales slumped again in 1992, despite a 25th Anniversary Trans Am and the special Firehawk. The latter was a hot, high-priced rival to the ZR1 Corvette, with 350hp (261kW) and 390lb ft (528.8Nm) from its tuned 5.7-liter V8. Add in competition options like Recaro seats and a Simpson five-point harness, and the Firehawk could cost as much as $50,000.

OPPOSITE: 1979 Firebird Trans Am, now with four rectangular headlamps.

Fifth and Final

The fifth-generation Firebird/Trans Am appeared in 1993. Now with rounded 1990s styling, to the casual observer it was little different from any other sports 2+2 of the time. And as ever, Pontiac's pony was little more than a restyled Chevrolet Camaro. It shared the new body and its 350 V8 with that eternal arch-rival, the only styling differences being below the waistline. Still, there was no denying that it was substantially new, with only 10 per cent of its parts carried over from the old car. The 1990s Firebird was still a rear-wheel-drive V8-powered 2+2, and in that at least little had changed.

Under the skin, a whole range of safety equipment included ABS brakes, plus driver, and safety airbags. There were just two basic engines, a 160-hp (119-kW) 207-cu in (3.4-liter) V6 and a 280-hp (209-kW) 5.7-liter V8. But the high-performance Firehawk had proved so popular that this returned, now with a fuel-injected LT1 power unit of 300hp (224kW) and the choice of six-speed manual or four-speed automatic transmission. Customers could expect a 13.5-second quarter-mile, so if there had been any doubt that performance was back in fashion, here was the proof.

As ever, there were both Firebirds and Trans Ams; for 1994 the former came with the V6 or a 275-hp (205-kW) version of the LT1, as base or Formula. Both Trans Ams (base and GT) were fitted with the 275-hp LT1 and six-speed manual transmission. The long tradition of performance options was upheld by the WS6 package: this tag had first appeared back in 1978 as a handling package, and it offered the same option in the mid-1990s, with larger wheels and tires, stiffer suspension, larger stabilizer bars, and four-wheel disc brakes. By 1996, the LT1 V8 had been boosted to 305hp (227.5kW) at 5,400rpm, which according to *Motor Trend* allowed the WS6-equipped Trans Am to accelerate to 60mph (96.5km/h) in

5.7 seconds and make the quarter-mile in 14 seconds dead. In short, the latest Trans Am was just as quick as the old ones, and complaints that the glory days of performance were over were simply wrong. That was underlined in 1998, when the LT1 was uprated again, this time to 320hp (238.5kW) and 345lb ft (467.8Nm), enough for a 0–60mph time of 5.1 seconds.

The following year was the Trans Am's 30th anniversary, and once again Pontiac made the most of its heritage. A limited run of both coupés and convertibles left the production line in white, with two blue racing stripes, just like the 1969 original. There were white leather seats and blue-tinted 17-in (3.2-cm) wheels. Meanwhile, the Firehawk remained the performance flagship. It was still built by SLP Engineering (which had produced the original Firehawk a decade earlier). By 2002, this was offering 345hp (257kW) and 350lb ft (474.6Nm) from its LS1 V8, 17-in Firestone low-profile tires, spoilers, and 9-in (22.86-cm) aluminium wheels. The Firehawk was not actually a model in its own right, but an option package that could be applied to a Firebird Formula, or the Trans Am coupé or convertible. By the end of 2002, Pontiac dropped all three; the Firebird, Trans Am, and Firehawk.

THE HEMI ENGINE

When one thinks of American muscle cars of the late 1960s and early 1970s, or indeed of any time, the one engine that stands head and shoulders above the rest is the Chrysler Hemi. It was quite simply the most powerful production engine of its time, so powerful, in fact, that Chrysler understated its real power by a substantial amount. In print, the Hemi produced a claimed 425hp (317kW) at 5,000rpm. In practice, it was more like 500hp (373kW) at 6,000rpm.

OPPOSITE: Dodge's Charger had the Hemi treatment.

ABOVE: The Hemi-equipped Plymouth Belvedere added up to a practical performance car.

But despite offering more power than anyone else, even at that "official" figure, during an age in which horsepower was king, the Hemi sold only in tiny numbers. It was an option on several Chrysler muscle cars, but there were few takers. In the five years that it was available as a road-going option, only 11,000 Hemis were sold, a miniscule number by U.S. mass-production standards. Maybe history is

playing tricks on us, and the legend that has grown up around the Hemi in the 30-odd years since, its domination of drag racing and NASCAR, has given it more prominence than it deserves. Maybe at the time a stock 383-cu in (6.28-liter) engine was quick enough, and a 426-cu in (7-liter) Hemi simply did not seem worth its higher insurance rating. What was uppermost in the minds of most buyers was probably the option price, anything between $500 and $1,100 or more, which was quite a chunk on a $2,700 car. Whatever the reasons for its limited sales, it has left us with quite a legacy.

Chrysler, in particular, had much to thank the Hemi for, because its effect on the public's perception of Chrysler muscle cars was out of all proportion to the numbers actually made. A clue to its secret lay in that name: Hemi was short for Hemispherical, indicating the use of a half-sphere-shaped combustion chamber. It is now recognized that this shape allows more room for larger ports and valves, and a higher compression ratio in relation to the size of the combustion chamber. With higher volumetric efficiency than a comparable engine, a hemi breathes deep, and deep breathing is one of the holy grails in the search for high horsepower.

But the 426 Hemi of the 1960s was not Chrysler's first such engine. A smaller 331-cu in (5.42-liter) engine was launched in 1951. The Firepower V8, as it was called, produced 180hp (134kW), though maybe it was wasted in the heavyweight Saratoga sedan: at 4,000lb (1814kg) that tended to blunt the Firepower's performance somewhat. A smaller 241-cu in (3.95-liter) version followed

OPPOSITE: 1967 Dodge Coronet R/T.

ABOVE: The Hemi fits neatly into the Coronet engine bay.

in 1953, fitted to Dodges, and in 1955 the new Chrysler C300 made a huge impact with its 300-hp (224-kW) 331-cu in (5.42-liter) Hemi. At the time, this was a stratospheric figure for a production car, and Dodge backed it up with a 270-cu in (4.42-liter) unit rated at 193hp (144kW) in Super Red Ram form. There was also an even higher-performance D-500 from Dodge, with more cubes and horsepower.

Chrysler dropped its first-generation Hemi in 1959, replacing it with a 413-cu in (6.77-liter) V8 with wedge-shaped combustion chambers. Nicknamed Max Wedge, it proved a formidable performer, especially once the race-tuned version had appeared in the early 1960s, as the Dodge Ramcharger or Plymouth Super Stock. With 11.0:1 compression, twin four-barrel carburettors, and an aluminium short-ram inlet manifold, the Max Wedge was soon making a big impression on the drag strip. Officially, it was designed for "police pursuit work:" in other words the drag strip! There was also a hotter 420-hp (313-kW) version with 13.5:1 compression. In 1963 the Max Wedge was taken out to 426 cu in (7.0 liters), which meant 425hp (317kW) in high-compression form.

A Racing Start

The Max Wedge was merely a preamble to the main act, the Hemi. Unveiled in 1964, it owed much to both the original Hemi and the Max Wedge, combining the deep breathing of one with the sheer cubic capacity and wild tuning of the other, but was really an all-new engine. At first, this was a pure race unit, with a compression of 12.5:1 and double-roller timing chain. Carburation was a single four-barrel carburettor for stock-car racing, or two four-barrels on the drag strip, the latter version rated at that infamous 425hp, with a true figure of over 500hp. Installed in radically lightened B-body Chryslers, the Hemi soon made a name for itself, especially in drag racing, where it rapidly achieved near-complete domination. The engine itself was lightened with aluminium heads in 1965.

So before it set foot anywhere near a road car, the Hemi was already well known by the driving public and, crucially, the sort of public that bought hot cars. It was only a matter of time before the two were put together, when for 1966 Chrysler announced the street Hemi as an option for certain models. It was obviously detuned from racing spec. The compression was lowered to a more moderate 10.25:1, there was a milder hydraulic lift cam and twin four-barrel Carter carburettors. Quoted power was still 425hp at 5,000rpm, with 490lb ft (664.4Nm) at 4,000rpm.

The Hemi road era had begun, but it was a low-key start. One could hardly miss the flashy Pontiac GTO or the imposing Ford Fairlane, for these were muscle cars with presence. But the Dodge Coronet and Plymouth Belvedere/Satellite were inoffensive-looking sedans. They were the sort of thing retired insurance salesmen would drive, or maybe a great aunt, as likely to rip away from the traffic lights in a plume of tire smoke as they would fly to the moon. There was a "426 Hemi" badge, but blink and one missed it, it was so small. But despite their dowdy exteriors, these were the fastest, most powerful cars on the market in 1966. Moreover, looks certainly did not bother *Car and Driver*, which tested a Satellite Hemi in 1966. This just missed a test of six "Super Cars" the previous month, but *C&D* made no bones about what the outcome would have been if the Hemi had made it in time. "Without cheating, without expensive NASCAR mechanics, without towing, or trailing the Plymouth to the test track, it went faster, rode better, stopped better, and caused fewer problems than all six of the cars tested last month." For the record, it dashed off 0–60mph (96.5km/h) in 7.4 seconds, with a time of 14.5 seconds for the standing quarter-mile. And there was another factor: despite its racing origins, the Hemi proved as reliable and docile a street engine as one could hope to find. Far from being trailered to the test (and for magazine tests of the time, some muscle cars were) the Satellite was driven from Detroit to New York, and was used every

OPPOSITE: There is no mistaking the performance potential of this 1968 Charger.

ABOVE: 1968 Dodge Charger R/T.

day of the week before the test. About the only complaint *C&D* could find to say about the fastest muscle car ever was its styling.

If, on the other hand, one really considered the Coronet and Belvedere too staid, one could always opt for the Charger, a fastbacked version of the Coronet which actually looked quite

different, thanks to the fastback's almost wedge shape and full-width grille with concealed headlights. The little doors that hid the lights were powered by electric motors. Underlining the upmarket approach, there were four bucket seats and full instrumentation. The Hemi was, of course, optional. As part of the Hemi package, the buyer received heavy-duty suspension, larger brakes, and 7.75 x 14 Blue Streak tires. Transmission was a four-speed manual or

ABOVE: The 1969 Charger R/T. Concealed headlamps were in vogue at the time.

OPPOSITE: 1969 Dodge Super Bee.

TorqueFlite automatic. However, of 37,000-plus Charger customers in 1966, only 468 (a little over 1 percent) paid extra for the Hemi package, split roughly 50/50 between TorqueFlite and four-speed. But over 1,500 Belvedere/Satellite buyers went for the Hemi, so maybe there were a few drivers out there who appreciated the combination of conservative looks with stunning performance.

As for engine longevity, Chrysler played safe with a reduced 12,000-mile (19310-km)/12-month warranty on the Hemi, and even this would be invalidated by what they euphemistically called "extreme operation." Still, it was a sensible move. With a car this fast, it seemed almost inevitable that keen owners would take it racing at weekends, before turning up at the dealer come Monday morning, with some broken parts and a warranty claim form in hand. They played even safer with the Hemi cars built for Super Stock: these had no warranty at all. A notice under the

bonnet reminded owners of the fact: "NOTICE: This car is equipped with a 426-cu in engine (and other special equipment). This car is intended for use in supervised acceleration trials and is not intended for trials or general passenger car use. Accordingly, THIS VEHICLE IS SOLD 'AS IS' and the 12-month or 12,000-mile vehicle warranty coverage and 5-year or 50,000-mile power train warranty coverage does not apply to this vehicle."

For 1967, Chrysler attempted to catch up with the flashier muscle car competition. The GTX was really a Belvedere with a massive hood scoop, stripes, rally wheels, and bucket seats. Although basically a Belvedere, it looked the part of a muscle car, and in Super Commando form came with the biceps to back it up, the choice of 440-cu in (7.21-liter) Max Wedge or Street Hemi V8s. Although smaller than the Max Wedge, the Hemi offered an extra 50hp (37kW) for an extra $564, which included heavy-duty suspension. So it was good value, but again, few GTX customers opted for it: 720 out of around 12,500 (and at least one author puts the figure at just 125; but, either way, the Hemi was in a minority).

Alongside the GTX, Dodge launched the R/T (Road & Track), a rebadged version of the same car with slightly tweaked styling. But neither was the fully styled muscle car that the opposition was offering. From Chrysler, that was still to come.

Chrysler may have come late to the muscle car scene, but lost little time getting in the swing, though of course with engine options like the Hemi and Max Wedge, that was not too much of a problem. Muscle cars were getting increasingly expensive by the late 1960s, and out of the reach of younger buyers, so Mopar led the field with the first budget muscle cars, or what *Car and Driver* magazine called "Econo-Racers." It was a simple formula: take the lightest, cheapest two-door body available, strip off all the options, and stick in the most powerful off-the-shelf V8. For car fanatics who loved trawling through the options lists this was no big deal, but the new Plymouth Road Runner did all that for one, offering a ready-made Q-car at a bargain price.

The Road Runner (and its Dodge Super Bee equivalent) was an instant hit, selling well over 40,000 cars in its first year. That made up nearly one in five of all intermediate Plymouths, quite a sales feat. It certainly did not look like a muscle car, lacking chrome, bulges, and hood scoops. Inside, rubber mats, rather than carpets, covered the floor and there was a plain bench seat: bucket seats were not obtainable, even as an option. Only the discreet Road Runner badges hinted that there was something special under the bonnet. That of course, was the Road Runner's raison d'être. The standard power unit was a special 335-hp (250-kW) version of Chrysler's 383, but more interesting by far was the $700 Hemi option. For this amount, Hemi buyers also had a larger radiator, power front disc brakes, and 15-in (45.7-cm) wheels with F70 Polyglas tires.

Dodge should not be forgotten, of course; its equivalent to the Road Runner was the Super Bee, based on the Coronet and aimed at the same concept of muscle car performance in a more affordable package. That was made clear in this gem of the 1968 copywriter's art: "Announcing: Coronet "Super Bee." Scat Pack performance at a new low price. Beware the hot-cammed, four-barrelled 383 mill in the light coupé body. Beware the muscled hood, the snick of the close-coupled four-speed, the surefootedness of the Red Lines, Rallye-rated springs and shocks, sway bar, and competent eleven-inch drums. Beware the Super Bee. Proof that you can't tell a runner by the size of his bankroll." As with the Road Runner, a 383 V8 was standard, but a Hemi topped the options list. Without the Hemi, the two-door sedan cost less than $3,500, but 166 customers chose to pay the extra $712. Actually, the acceleration figures were not much different: the standard Super Bee reached 60mph (96.5km/h) in 6.8 seconds, and the Hemi was just 0.2 seconds quicker, though it did lop a whole second off the quarter-mile time, at 14 seconds. For 1971, with the Coronet out of production, the value-for-money Super Bee option was available on the Dodge Charger only.

OPPOSITE: For 1970, the Hemi finally appeared in the Chrysler pony cars, the smaller, lighter Dodge Challenger (shown here), or the Plymouth Barracuda.

Meanwhile, Road Runner sales nearly doubled to 80,000 in 1969, and *Motor Trend* named it Car of the Year. Not many of those 80,000 were Hemi-powered, partly because a new triple two-barrel carburettor option was added to the 440-cu in (7.21-liter) V8, which offered Hemi style acceleration at half the price. But for the full experience one had to have a Hemi, a fact underlined by *Car and*

A 1970 Lemon Twist Challenger convertible. Other colors offered by Chrysler were Vitamin C, Lime Light, and Go-Mango. Sunglasses were optional.

Driver when it tested the Hemi Road Runner against five other econo-racers, the Chevrolet Chevelle SS396, Ford Cobra, Mercury Cyclone CJ, Dodge Super Bee, and Pontiac The Judge. Put simply, it was in a different class. The Hemi option made this Road Runner several hundred dollars more expensive than any of the others, but if the following statement is correct, it was worth it: "Where the Chevelle, Cobra, and Cyclone CJ give the

ABOVE: 1968 Plymouth GTX convertible.

LEFT: The 440 Super Commando engine option was a little slower than a Hemi, but much cheaper.

impression of being hot sedans, the Road Runner comes in from another direction – a tamed race car… [At full throttle] the exhaust explodes like Krakatoa and the wailing howl of surprised air being sucked into intakes turns heads for blocks. Baby, you know you're in the presence." In short, the Road Runner accelerated faster than any rival (5.1 seconds to 60mph and a 13.54-second quarter-mile), stopped more quickly and rated second out of the six on handling. They also warned about "incredibly stiff" suspension. And yes, it was worth it.

In the same year that *Car and Driver* was eulogizing the Road Runner, Dodge announced something far more outrageous, the Charger Daytona. From its 1966 luxury fastback origins, the Charger had gradually become more and more of a muscle car. Sales slumped in 1967, so for 1968 it was completely restyled to distance it from the Coronet on which it was based. There was a coke-bottle kick-up to the side panel, flared wheel arches, and more aggressive fastback shape. The Hemi option came only with the R/T version of the Charger, and in 1968 475 were sold, a figure that had dropped to just 42 by 1970.

The new Charger's high-speed performance was hampered by that deep-set grille and tunneled rear window, so for 1969 the Charger 500 was launched with a smoother, more aerodynamic shape in loud colors, big stripes and a 440 base engine option. The "500" came about as a result of NASCAR rules, which stated that 500 production-line cars had to be built to qualify for racing. But meanwhile Dodge's main NASCAR rivals, Ford and Mercury, had equalled the 500's aerodynamic improvements. So Chrysler spent an alleged $1 million on a car specifically developed for NASCAR. It was developed in a wind tunnel, the wedge-shaped nose adding a whole 18in (45.7cm) to the car's length. At the rear, there was a massive spoiler, 58-in (1.47-m) wide and nearly 2ft (61cm) clear of the boot lid. It looked like the ultimate street racer's weekend toy, but really did reduce lift at high

LEFT: The Road Runner lived up to its cartoon inspiration.

BELOW: 1969 Plymouth Road Runner.

speeds. As a result of all these changes, drag was cut by 15 percent and in race trim the Charger Daytona could reach 200mph (322km/h). It won first time out, too, at Talladega. Bobby Isaac set a world closed-course speed record in one at 201.104mph (323.64km/h), and went on to set an unlimited class record at Bonneville Salt Flats, with 217mph (349km/h).

Outrageous though they were, the Daytonas really could be road cars. Five hundred of them had to be, to qualify for NASCAR racing, and the first was delivered to a Dodge dealer in Lafayette, Indiana in June 1969, the last reaching the saleroom in September. Most were powered by the base 440 engine, but 70 Daytonas came Hemi-powered. Interestingly, the rules were changed in 1970. Instead of 500 cars to qualify, manufacturers had to build one for each dealer. Plymouth produced its own version of the Daytona, the Superbird, in that year, and so nearly 2,000 were built, just 135 of which had the Hemi option. Although it looked very similar to the Daytona, with the same droop snoot and high rear spoiler, the Superbird was quite different, being based on the Road Runner. In the early 21st century Chrysler's "Winged Warriors," as they were known, look faintly ludicrous. One can imagine them being shown on the Thunderbirds TV show, or driven by Batman on his day off. But that is unfair: as road cars they were outrageous, but behind that was a list of race wins that justified the outlandish looks – they weren't for show.

The Pony Hemis

But times were changing in the muscle car world. The "Winged Warriors" were a symbolic last flamboyant gesture in a world where sheer brute power was going out of fashion. Emissions and safety legislation was getting tighter by the year, insurance premiums were going through the roof, and there was a growing climate of public

opinion opposed to what was seen as a profligate use of resources. In truth, it was the end of the pure, no-holds-barred muscle car era.

But there is another piece of the Hemi story still to be told. Cars like the Charger and GTX, even the Road Runner, were too big and heavy to compete with the pony cars, the immensely successful Mustang and Camaro. For that market, Plymouth had launched an all-new Barracuda in 1966, in fastback, coupé, and convertible forms. But it was not until 1970 that a Hemi option was added: the mighty 426 was simply too big to squeeze in before that, even after the Barracuda's 1968 facelift.

In that year Chrysler contented itself with building some lightweight Dodge Darts and Plymouth Barracudas for drag racing. The front spring towers were prized apart to allow the big Hemi to be squeezed in. But a purely road-going Barracuda/Dart had to wait for the 1970 E-body, which was designed specifically to allow bigger engines such as the Hemi and 440. In the words of author Robert Genat, "Chrysler was determined to build the most potent pony car ever." To handle that power, the cowl, front subframe and rear axle were taken from the B-body cars (the Coronet/Charger) and a new muscular body was designed to suit, lower and wider than the old Barracuda, and in the classic long hood/short trunk mould. There would be both a Plymouth Barracuda and an upmarket Dodge Challenger, both based on the same platform, but with subtly different bodywork. Flush door handles, hidden wipers, and streamlined mirrors were all period touches. Fat 60-series Polyglas tyres emphasized the aggressive stance. What it did not have was a urethane front end like the Pontiac GTO. Chrysler did not have the budget, but offered the Elastomeric, a steel bumper covered in high-density foam, painted in body color.

Once one had decided to buy a Barracuda or Challenger, this was just the beginning. There was a base Barracuda, a luxury Gran Coupé, and sporty 'Cuda, the latter with a Hemi option. Challenger trim levels covered base, SE and R/T. The Barracuda engine range was huge, starting with two non-muscular slant sixes (125 and 145hp/93 and 108kW) plus a 318-cu in (5.2l-liter) V8. Then one stepped up to a 275-hp (205-kW) four-barrel 340-cu in (5.57-liter) unit, and the 383-cu in (6.28-liter) engine in 275-hp (205-kW) two-barrel and 300-hp (224-kW) four-barrel forms. Topping out the non-Hemi range was the "Six-Pack" 440-cu in (7.2l-liter) unit that produced as much torque as a Hemi but a "mere" 385hp (287kW). But for true performance freaks with deep pockets and thick wallets, there could only be one choice, the Hemi. Choose the latter for one's 'Cuda, and it came as standard with the famous Shaker scoop, that burst through the hood and lived up to its name, trembling and quaking in sympathy with the V8 to which it was bolted. The Hemi option was one of the most expensive, at $1,227, which added over 40 per cent to the price of the basic car. Despite this penalty, 666 customers ordered it, making the 'Cuda the second most popular Hemi-powered Chrysler. They would not have been disappointed either, as *Motor Trend* tested one in September 1969, with TorqueFlite transmission, 4.1:1 rear axle and power steering. Despite considerable wheel spin, it tripped the quarter-mile clocks at 13.7 seconds and 101.2mph (162.9km/h).

The Hemi 'Cuda option may have been short-lived and rare, but it achieved fame a quarter of a century after it disappeared from the Dodge price lists. Don Johnson, who had played a Ferrari-driving cop in the TV show *Miami Vice*, wanted a Hemi 'Cuda convertible for his new cop show series, and called in Hollywood car supplier Frank Bennetti. The trouble was, Hemi 'Cudas are rare, expensive collectors' items. Even if they could find a car in time, explained Bennetti, it would cost a fortune. So in true Hollywood style, they cheated. Four Barracuda convertibles were painted an identical yellow, trimmed with a correct 1971 Parchment white interior, and went filming. Under the bonnets were 360-cu in (5.9-liter) or 440-cu in (7.2l-litre) V8s, far less valuable than the real thing. But as far as

the viewing public was concerned, or those who knew their muscle cars, it was fame at last for the Hemi-powered 'Cuda.

As for the Challenger, one could have the Hemi option only in sporting R/T trim, though in any of the body styles: hardtop, SE hardtop (a special for 1970), and convertible. As ever, it was the most expensive engine option. In place of the standard 383, one shelled out an extra $130 for a 375-hp (279.5-kW) 440-cu in (7.21-liter) Magnum V8, or $249 for the hotter 390-hp (291-kW) version. The Hemi was in a different league, adding nearly $780 to the basic price. The Challenger was a big hit for Dodge, in 1970 alone selling 76,000, of which just under 20,000 were

R/Ts. Of those 20,000-odd, there were only 356 Hemi-equipped cars. Not that Dodge would have been too bothered: the Challenger outsold its arch rival, the Mercury Cougar, that year, which was sweet for Chrysler.

Sweet perhaps it was, but it was only a brief moment of glory. In 1971 all muscle car sales slumped, because of the pressures mentioned above. Naturally, all-out performance options like the Hemi suffered even more. In that year Chrysler sold just 115 Hemi-equipped 'Cudas and 71 Challengers. By now, the only B-models available as Hemis were the GTX and Road Runner: in 1971 Plymouth sold just 55 Hemi Road Runners and 35 GTXs. Nothing more clearly spot-lighted the fact that the age of the full-horsepower muscle car was drawing rapidly to a close. And let there be no doubt, a Plymouth or Dodge with a 500-hp (373-kW) race-bred V8 under the bonnet was certainly one of those.

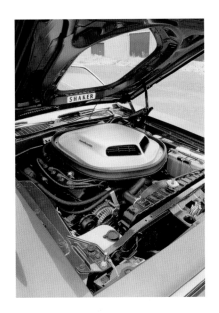

OPPOSITE
1976 Plymouth 'Cuda Hemi.

ABOVE
The Hemi option was powerful but expensive. The Shaker scoop was standard.

OTHER CLASSIC MUSCLE CARS

GTO, Mustang, Hemi, Trans Am, Firebird: these are the muscle cars everyone knows, but there are plenty of others. The 'Big Three', plus American Motors, all produced a great deal of muscular metal in the 1960s and 1970s. Some were hardly worthy of the name, and most were mildly pepped-up versions of existing cars; but all had performance as a primary aim. This chapter does not attempt to cover all of these, for that would fill an entire book; it aims simply to describe a representative selection.

BELOW & OPPOSITE: American Motors'
1969 AMX two-seater was smaller
than other muscle cars.

OPPOSITE & BELOW: Although the AMX interior provided ample space for two, there was limited space for luggage so a trunk rack was provided.

AMERICAN MOTORS: RAMBLER & AMX

It had never been easy for American Motors to compete with the might of Ford, General Motors, and Chrysler, and yet it survived the 1960s and 1970s, only succumbing in the end to a takeover by Chrysler in 1987. It never produced a rip-roaring muscle car in the classic tire-smoking tradition, but neither was it

OPPOSITE & BELOW: The Mid-engined AMX/3 of 1970.

completely left behind. AMC's first contribution to the muscle era arrived in 1966 with the Rambler Rogue V8. There had been a 270-hp (201-kW) fastback Marlin the year before, but that was limited by AMC's venerable V8 engine, which could not be stretched beyond 327 cu in (5.36 liters).

The Rogue used the new Typhoon motor, initially as 290 cu in (4.75 liters) in two-barrel 200-hp (149-kW) or four-barrel 225-hp (168-kW) forms. This was far more stretchable, and was eventually expanded right up to 401 cu in (6.57 liters). The Rambler was, of course, a compact by American standards, so the Typhoon gave it respectable if not tarmac-melting performance; in fitting it AMC was following the lead of Ford and Chevrolet, which had both shoehorned V8s into their own compacts.

The Typhoon was also slotted into the bigger Marlin the following year, this time in 280-hp (209-kW) 343-cu in (5.62-liter) guise. But the Marlin was no lightweight, and took nearly 10 seconds to reach 60mph (96.5km/h). Better was to come, however. The attractive fastback AMX, which resembled a 2+2 but was actually a pure two-seater, turned out to be AMC's first true muscle car. The 390-cu in (6.39-

liter) Typhoon put out a useful 315hp (235kW), which turned out to be more than useful when powering the little AMX, which itself weighed only a little over 3,200lb (1452kg). *Car and Driver* recorded a time of 6.6 seconds for 0–60mph (0–96.5km/h) and 14.8 seconds over the quarter-mile, estimating top speed at 122mph (196km/h). AMC had been disadvantaged by a somewhat staid and sensible image, but the AMX marked a complete break with that, and it succeeded. It gave AMC a toehold (albeit a small one) in the youth market, especially after Craig Breedlove established 106 new world records with the AMX early in 1968.

OPPOSITE, RIGHT & OVERLEAF: From 1971, both AMX and Javelin shared the same restyled body, now with a 401-cu in (6.6-liter) engine option.

The Javelin, launched in 1968, looked very similar but had a much longer wheelbase, four seats, and the Mustang in its sights. Like other pony cars, it came with a wide range of motors, kicking off with a 232-cu in (3.8-liter) six, but the real muscle variant was the SST, with that 390 that worked so well in the lighter AMX. This option was introduced late in the 1968 model year, so late that there was no mention of it in AMC brochures! At first, it was hampered by a wide-ratio three-speed manual gearbox, but a four-speed 343 Javelin was almost as fast. But for 1969, it came into its own, especially with the loud and proud options aimed directly at the muscle car aficionados. The "Go-Package" brought heavy-duty suspension and wide wheels, or one could choose from garish new colors: Big Bad Orange and Big Bad Green. If a standard Javelin 390 was not quick enough, there were tuning parts down at the dealers, and a whole variety of rear axle ratios. There could be no doubt about it, AMC was determined to leave its sensible past behind,

spurred on by an encouraging start to its racing program.

It took things one step further in 1969, teaming up with tuning specialist Hurst to produce the hottest Rambler yet. It was actually Hurst which had the idea of combining the lightest 3,000-lb (1361-kg) Rambler with the 390-cu in (6.4-liter) V8 in 325-hp (242-kW) form. That gave it the terrific power to weight ratio of 10.03lb/hp (6.1kg/kW); according to Road Test magazine, when combined with the four-speed Borg-Warner box, it produced a quarter-mile time of 14.4 seconds at just over 100mph (161km/h). No one could mistake the SC/Rambler for any lesser AMC, with its loud red, white, and blue color scheme, wide wheels, and red-stripe Goodyear tires. Once again the SC proved that the classic muscle formula of big engine in a small light car was a simple way to high performance. The same 325-hp 390 found its way into the AMX and Javelin SST as well, and these too were quick cars. AMC also collaborated with Hurst in 1970 with the "Rebel Machine,"

now with 340hp (253.5kW) from the faithful 390, with Ram Air induction and a low back pressure dual exhaust. This too came with an extrovert red, white, and blue paint job, but if that was too much, one could go for something more sober. In the event, it was these 390s which proved to be the ultimate AMC muscle cars.

For 1971, the AMX and Javelin benefited from the company's biggest, most powerful engine yet, the 401-cu in (6.57-liter) V8 with 330hp (246kW). The trouble was that they now both shared a bigger, heavier bodyshell, so the power/weight ratio was actually down, though it was still a fast car, at 7.7 seconds to 60mph (96.5km/h) for the AMX, and a 15.5-second standing quarter-mile. By now, muscle cars were attracting high insurance costs, and AMC responded with a mild-muscled version of the compact Hornet, which had replaced the Rambler. The 360-cu in (5.9-liter) V8 was not AMC's biggest or hairiest, but it was enough to propel the lightweight Hornet over the quarter-mile in 14.8 seconds. In a few short years, these modest muscle cars had helped AMC to transform its image.

OPPOSITE: 1971 Javelin interior.

BUICK GRAN SPORT

William Mitchell, so the story goes, wanted to create an American Jaguar. As General Motors's head of styling, he liked the way Jaguars combined sportiness with luxury in a sedan package. The new Riviera of 1963 was not that, but the Gran Sport derivative, which appeared a couple of years later, was much closer. In contrast to other Buicks and, for that matter, most other American cars, the Gran

OPPOSITE: 1963 Buick Riviera.

BELOW: 1962 Buick Riviera V8.

Sport came with just one engine and transmission option. But it did not need alternatives as the 425-cu in (6.96-liter) V8, with its 10.25:1 compression, produced 360hp (268.5kW), and a massive 465lb ft (630.5Nm) of torque.

According to *Car and Driver*, its three-speed automatic was "without question, the best automatic transmission in the world."

Twin four-barrel Carter carburettors and big dual exhausts pointed to the power ambitions of this Detroit Jag. There were suspension changes too: a front anti-roll bar and stiffer springs and shocks all round. All this, and a time of 7.2 seconds for 0–60mph (0.96.5km/h), made the Gran Sport a fully-fledged muscle car.

In that same year of 1965 Buick unveiled the Skylark GS (Gran Sport). This was a direct response to Pontiac's GTO, being smaller than the Riviera. In fact, "direct response" is quite an appropriate: an early GS advertisement described the car as "your own personal-type nuclear deterrent!" It was not quite as forceful as the big-engined Riviera, but the 401-cu in (6.57-liter) Wildcat

1968 Riviera Gran Sport 430/4.

BELOW, OPPOSITE LEFT & OVERLEAF:
The GSX was very different from Buick's
earlier models and seems to have
abandoned all notions of subtlety.

OPPOSITE RIGHT: The 455 was the
biggest Buick engine.

power unit allowed 325hp (242kW)
and 0–60mph (0-96.5km/h) in 7.8
seconds. It was also held back by a
two-speed automatic and just one four-
barrel carburettor.

GS buyers had a choice of six different
rear axle ratios, however, and the car
could be coaxed into impressive drag
strip performance. *Motor Trend* tested

one with a 4.3:1 rear end, racing slicks, headers, shimmed front springs, and a transmission kick-down switch. It responded with a standing quarter-mile of 14 seconds at 101mph (162.5km/h), reaching 60mph in 5.7 seconds.

Well aware of the growing interest in muscle cars, Buick upped the GS's power with a 340-hp (253.5-kW) option introduced later in the year: even without special tires and ratios, that could trip the lights at just under 15 seconds. In fact, the muscle trend was spreading right across the division, and in 1966 Buick offered a GS version of the full-sized Wildcat, complete with heavy-duty suspension, Positraction, and the 340-hp V8.

Meanwhile, the Skylark GS range was expanded with the cheaper 340, with a 340-cu in (5.57-liter) motor giving 260hp (194kW) at 4,200rpm and a useful 365lb ft (494.9Nm) at 2,800rpm. The engine was a common fitment to mid-sized Buicks at the time, but in the GS 340 came with a four-speed manual gearbox, red stripes, badges, and hood scoops to leave no one in any doubt that this was a budget

muscle car. In the event, most buyers (4
to 1) went for the GS 400, and the 340
was soon replaced by the slightly beefier
GS 350.

For buyers of muscle cars, the 400 was
still the Buick to have, especially in 1967,
when it received a brand-new 400-cu in
(6.555-liter) engine. Of lighter weight and
with better breathing than the old Wildcat
motor, it was slightly oversquare, though
the claimed power was no different from
that of the 340-hp Wildcat. To match it,
GS 400 buyers (the car was now a model
in its own right) could opt for the latest
variable-pitch-stator Super Turbine
transmission for an extra $237, which
until then had only been available on the
big Buicks. But the four-speed manual
option was still the fastest: it was a full
second quicker to 60mph (96.5km/h),
for example.

Reflecting these muscular times, the
400 was getting gaudier, with more
stripes, side vents, and hood scoops, and
one could pay $90 extra for chrome-

plated wheels. Its performance was getting more serious too. For 1969, there were two optional levels of tune. Stage I was dealer-fitted, a comprehensive list of parts that included a high-lift cam, high-output oil pump, heavy-duty valve springs, and tubular (thus lighter) pushrods. A large dual exhaust and modified Quadrajet carburettor were part of the package, as was a 5,200rpm governor in the transmission to prevent over-revving, and a choice of 3.64 or 3.42:1 Positraction rear axles. Pay a little more, and one could have heavy-duty suspension and power front disc brakes as well. Fitted up and ready to go, the GS Stage I produced 345hp (257-kW) at 4,800rpm. Stage II was intended only for racing, and was not recommended for use on the street or for any car with a silencer. The Buick dealer was not even able to fit the parts: they had to be bought over the counter, from which point the buyer was on his own. That year, a GS 400 was the fastest car tested by *Car Life* magazine, and achieved 0–60mph in 6.1 seconds.

But cubic inches were the final arbiter in the muscle car wars, and Buick obliged in 1970 with the Gran Sport 455. This latest 455-cu in (7.46-liter) V8 was a stretched version of the 430, and produced 350hp (261kW) at 4,600rpm, breaking the 500-lb ft (678-Nm) barrier too, at 510lb ft (691.6Nm) at 2,800rpm. That was in a relatively mild state of tune (single four-barrel carburettor and 10.0:1 compression), but once again Stage I was ready and waiting. For a little under $200, buyers could have extra-large valves, big-port cylinder heads, stronger valve springs, a high-lift cam, modified carburettor, and even blue-printed pistons. Rallye Ride stiffer suspension cost just $15.80 extra: Buick seemed determined to give the horsepower fanatics a good deal! They would not have been disappointed either, as the Stage I GS could rocket to 60mph (96.5km/h) in 5.5 seconds and over the quarter-mile in less than 14 seconds. Not surprisingly, these Stage I 455s were popular, making up more than 1 in 4 Buick sales in 1970.

But all of these GS Buicks, however impressive their performance, still looked quite subtle, even understated, compared with the more garish muscle cars, but Buick responded with the GSX; bright yellow with rear spoiler, black hood, and rally stripes, this was really a GS (either with the 350-hp/261-kW V8 or Stage I) in a flashy new suit, plus all the performance options: hood-mounted tachometer, power front disc brakes, four-speed manual box, bucket seats, stiffer suspension – the list went on. Less than 700 GSXs were sold in 1970, suggesting that Buick's traditional clientele preferred the classy look, but that did not prevent the company from offering the GSX package on any Gran Sport in 1971–72 as well.

The Gran Sports were tamed and detuned in the early 1970s, but 1982 saw a new generation of Buick muscle cars. The Regal Grand National (to commemorate Buick's NASCAR win) was actually built for Buick by Cars & Concepts of Michigan. As standard, it came with a flabby 4.1-liter V6 of 125hp (93kW), but a few cars were fitted with the Regal Sport Coupé's turbocharged 232-cu in (3.8-liter) V6, boasting 175hp (130.5kW). The idea behind the turbo was to provide something closer to V8 power with V6 economy. The figures may have looked pale next to a late 1960s muscle car, though of course they were SAE net rather than gross; but by 1987 the V6 was intercooled and delivering 245hp (183kW) and 355lb ft (481.4Nm), powering a range of all-black Buicks.

OLDSMOBILE 442

If there was one thing that helped to keep General Motors healthy in the 1960s, it was the competition encouraged between its divisions. They shared some components, of course, and that made economic sense for all concerned, but as far as sales went they were left to sink or swim on their own. Apart from the odd directive from top brass (such as the banning of triple carburettors on any GM car except the Corvette) they were left strictly alone.

So as soon as Pontiac launched the GTO package, its sibling rivals lost little time in coming up with GTOs of their own. Buick had the Skylark Gran Sport, so what would Oldsmobile have? The answer was the 442. Those numbers would be as evocative to muscle car fans as GTO or even 409. Officially, they stood for four-barrel carburation, four-on-the-floor, and two dual exhausts. At least, they did until the four-speed manual gearbox could be replaced by an optional automatic, and

then Oldsmobile claimed that the second "4" represented cubic inches.

Like GTO, the 442 started out as an option package, in this case "Option B-09 Police Apprehender Pursuit," to give it its full official title. A $285 option on the Oldsmobile Cutlass, this brought 310hp (231kW), an increase of 20hp (14.9kW), as a result of a high-lift cam and high compression, that four-speed manual and dual exhaust. There were also heavy-duty shocks and springs and a rear stabilizer bar. As well as being as fast as a GTO, the 442 rapidly gained the reputation of being the best-handling muscle car of all, an accolade it retained right through the 1960s.

Word soon got round, and the 442 became a popular option, especially from 1965 when Oldsmobile shoehorned in a bigger 400-cu in (6.555-liter) engine: this

OPPOSITE & RIGHT: The Oldsmobile 442.

was a downsized version of Oldsmobile's new 425-cu in (6.96-liter) motor, and incidentally had the added benefit of more cubic inches than the 389-cu in (6.375-liter) unit in the GTO. In practice, it meant 345hp (257kW) at 4,800rpm and a hefty 440lb ft (596.6Nm) of torque, enough for a 0–60-mph (0-96.5-km/h) time of 5.5 seconds and a standing quarter-mile of exactly 15 seconds. *Car and Driver* loved it ("a very worthwhile balance of all the qualities we'd like to see incorporated in every American car") and so did the buying public: just over 25,000 of them ordered the 442 package that year.

OPPOSITE & ABOVE: The 442 (W30 apart) was never a stripped-down performance special, as this white-upholstered 1971 holiday hardtop shows.

For 1966 the Cutlass was restyled, while the 442 package now included convenience features like two-speed wipers and bucket seats. Another 5hp (3.7kW) was coaxed out of the V8, thanks to a higher compression, but the big performance news was an optional triple two-barrel carburettor set-up, which claimed 360hp (268.5kW). This was Tri-Power, but it was only available for a year before the GM management handed down the banning order. Still, the standard 350-hp (261-kW) 442 in 1967 was surely fast enough, especially when it cost a mere $184 extra. As well as the engine, it brought heavy-duty suspension, wheels and engine mounts, wide, low-profile tires, and those desirable little red "442" badges. If one insisted, the bucket seats could be swapped for a good old-fashioned bench seat. That year, the 442 option was available on all Cutlass Supreme two-door cars, which indicated its popularity.

But the 442 had become such a strong badge in its own right that, just like the GTO, it was made a model in its own right. For 1968, the 442 was still a hopped-up Cutlass, but was not badged as such. There were three of them: holiday hardtop, sports coupé, and convertible, all with a new 400-cu in (6.555-liter) engine in 325-, 350- and 360-hp (242-, 261- and 268.5-kW) forms. The last gave the same power as the triple-carburettor Tri-Power, but Oldsmobile got around the ban by finding the extra power in a different way. It was almost as if there was unspoken warfare between the top brass bureaucrats and the designers on the shop floor.

There was another way to avoid GM bans: team up with an outside supplier. That is what Oldsmobile did with famed tuner George Hurst. Hurst had already squeezed Oldsmobile's biggest 455-cu in (7.46-liter) motor into a 442, and this so impressed Oldsmobile that dealer John Demmer of Michigan was persuaded to build replicas, and being dealer specials, no GM power limitations could apply! This car was the first in a long line of Hurst/Oldsmobiles, which used the standard 442 base with a Force Air 455 supplying 390hp (291kW) and 500lb ft (678Nm). The engine itself was special, based on that of the Tornado but with special crank, high-lift/duration cam, Ram Air cylinder heads, and a modified auto transmission: this last was equipped with a Hurst Dual-Gate shifter that could be operated either manually or left to itself. It was quick, according to *Super Stock* magazine, to the tune of a 12.9-second quarter-mile, tripping the lights at 109mph (175km/h).

The hottest factory 442 remained the W30, which had started out as a drag-racing special but was now firmly aimed at the man in the street, though with optional rear axles down to 4.66:1, it is fair to say that many W30s ended up on the strip anyway. Little cosmetic touches underlined the 360-hp (268.5-kW) W30's status: bucket seats, red stripe tires, and hood stripes. There was also a slightly milder W31 package for 1969, with Turbo Hydra-Matic transmission, but this made up only one per cent of Oldsmobile's sales that year though. For 1970, the W30 package also included 10.88-in (27.64-cm) front discs, as well as the familiar Ram Air induction system, for which Oldsmobile claimed only an extra 5hp (3.7kW) over the standard 442. What everyone agreed was that the 442 was still the muscle car of choice for handling.

But times were changing, and 1971 was the last year for the 442 as a separate model: just like the GTO, changing times brought a decline in popularity, and the first sign that the manufacturer were aware of this was a reversion of these muscle car models to option packages, rather than being models in their own right.

By the end of the 1970s, the wild times of 360-hp (268.5-kW) 455-cu in (7.46-liter) 442s must have seemed a distant memory indeed, but in 1979 there was an attempt to inject a little of the old excitement back into the new downsized Cutlass. For around $2,000 (a lot of money on top of a $5,631 base price) buyers received a bigger 350-cu in (5.735-liter) V8 of 170hp (127kW) in place of the standard 130-hp (100-5-kW) 305 unit, with gold paint job, gold aluminium wheels, and a Hurst Dual-Gate shifter on the auto transmission. In sanitized form, the Hurst/Oldsmobile lived on. In fact, so strong was the name that it returned for 1984, complete with three-lever Lightning Rod Automatic Shifter, 180-hp (134-kW) 307-cu in (5.03-liter) V8, heavy-duty suspension and fat tires: it sold 3,500 in that year, and was the most popular Hurst/Oldsmobile ever.

CHEVROLET CHEVELLE

Just as Oldsmobile quickly responded to the GTO threat in 1964, so did Chevrolet, though at first it was turned down by General Motors's top management. The concept of a biggish V8 bolted into a mid-sized car had obviously worked so well for Pontiac that GM's "cheap car" division wanted to do the same. Oldsmobile had lost no time in turning the Cutlass, with police-specification engine and suspension, into the famed 442. Chevrolet, not unreasonably, decided to fit its small-block 327-cu in (5.36-liter) engine into the Chevelle. Inexplicably, the concept was turned down: the 327 was actually mild by comparison with the 310-hp (231-kW) Oldsmobile unit, let alone the GTO's 325-hp (242-kW) 389-cu in (6.375-liter) motor.

OPPOSITE: A 1969 Chevrolet Chevelle.

*BELOW: The 1967 Chevrolet Chevelle
SS396 convertible, combining the mid-size
Chevy with well over 300hp.*

Fortunately for Chevrolet, head office
relented, and the Chevelle SS 327 was
born. The Chevelle actually made a good
base for a muscle car. First of all it was
cheap, the lowest priced of GM's four A-
body mid-sized cars. There was already a
Super Sport (SS) option with bucket seats,

console, and badging, but with a maximum of 220hp (164kW) from the biggest V8 available it was no GTO scarer. Finally, the small-block 327 would slot straight in, and the engine itself was a proven quantity. So in the middle of that 1964 model year, two new options joined the Chevelle's list: the L30 (250-hp/186.5-kW) and L74 (300-hp/224-kW) versions of the 327, costing $95 and $138 respectively over the base 283-cu in (4.64-liter) V8. In theory, one could even have the Corvette's hot 365-hp (272-kW) version of this engine, the L76, but it is thought that these super-Chevelles went in limited quantities only to drag racers.

So the Chevelle had its muscle derivative at last, but it was not really strong enough to challenge a GTO or 442. For that it needed a big-block motor, and this was not long in coming. Not that it was just a case of slotting Chevrolet's 396-cu in (6.49-liter) engine straight into the Chevelle. In fact, it was so awkward that in its first year Chevrolet did not even advertise the fact that a 396 Chevelle was available, preferring to sell it in very limited numbers.

The problem was that to fit the big-block was expensive and time-consuming. A convertible frame was used with a coupé body, with two extra body mounts, and rear suspension reinforcements. Special left- and right-hand exhaust manifolds had to be made to squeeze the bigger engine in. Bigger power-assisted brakes (11-in/27.9-cm drums all round) were fitted, as were stiffer springs and shocks, plus stronger hubs, and wider wheels. Even the Chevelle's ring-gear had to be swapped for a bigger one, and the engineers also added a bigger 11-in (27.9-cm) clutch and four-speed Muncie gearbox. Just 201 of these Chevelle "Z16" SS396s were built for 1965, so it was not possible for a prospective purchaser to stroll into his local Chevrolet dealer, order one and expect immediate delivery. No matter, Chevrolet had made its intention clear.

This intention was made clearer still in the following year, when the rebodied Chevelle came as an SS396 from the start. In fact, there were three of them. The base 396 (it was a de-stroked version of the famous 409) produced 325hp (242kW) at 4,800rpm and 410lb ft (556Nm). If that was not enough, one could opt for the 360-hp (268.5-kW) L34. This had lots of internal changes: a higher-lift cam, chrome piston rings, forged alloy crankshaft, and dual exhaust, though it was only slightly torquier than the base L35, majoring on that substantial power increase. Still not enough? Then there was the top L78 with hotter-still cam, 11.0:1 compression and other parts to give 375hp (279.5kW) at 5,600rpm. That could reach 60mph (96.5km/h) in an alleged 6.5 seconds, but the L78 was in a very high state of tune, and it is thought that only around 100 Chevelles were equipped with it. The point was that a big-block Chevelle out-cubed the GTO and was now available at any Chevrolet dealer, in four-speed manual or automatic form, sport coupé or convertible. Together, over 72,000 of these were built in 1966, with another 63,000 in the following year.

The engine range was trimmed in 1967, though. The base engine was still the 325-hp L35, while the L34 was derated to 350hp (261kW) at 5,200rpm. As for the top-powered L78, this 375-hp unit was not available as a factory option any more, but one could buy all the necessary parts at a local dealer. Whatever the engine choice, a Chevelle 396 came with three- or four-speed manual gearbox, or a Powerglide automatic, with no less than nine rear axle ratios available, between 3.07:1 and 4.10:1. With the 350-hp motor and the 4.88:1 axle, one could turn in some impressive performances on the drag strip.

For 1969, the 375-hp option was back, a clear sign of the times that the muscle car war was hotting up: 350hp as top power just was not sufficient any more. The Chevelle had a new body that year, the popular long hood/short deck look, with a fastback on the coupé, though sales slid again to 57,000. For 1970, the SS396 reverted to option package status rather than being a model in its own right. Normally this is a signifier of low and falling sales, but that year they rocketed right

In 1969, over 86,000 Chevelles were ordered with the SS package, which included the 396-cu inch engine, wide wheels and tires and all sorts of badges and trim details.

back up to over 86,000, the best yet for the Chevelle SS.

Smart but not flashy, the package brought lots of detail items to differentiate the SS from lesser Chevelles: twin hood bulges, black grille, bright wheel rims,

The Chevelle was only one Chevy available with the SS package. By 1969 the Impala SS had grown into a big four-seater.

and roof drip moldings, rally stripes, and white-letter F70 x 14 tires with 7-in (17.8-cm) sports wheels. At $440 it was still good value, especially as the heart of the 396 was now the 350-hp engine, with a new 375-hp unit with aluminium heads

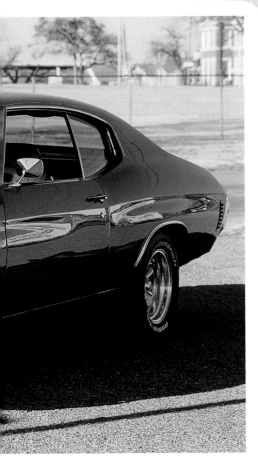

(and probably more power than was claimed) as the top option.

Or was it? Through GM's infamous Central Office Production Order (COPO) system, it was possible to order a Chevelle with even more power, direct from the factory. COPO (as detailed in the Camaro chapter) was intended as a means of satisfying fleet managers who wanted non-standard specification cars; for example, a bench seat in place of standard buckets, or a cheaper three-speed gearbox instead of a four-speeder. But if one knew the system, it could be used to order big-block engines. Chevrolet dealer Don Yenko did just that with the 427-cu in (7.0-liter) big-block V8, and some of these found their way into very special Chevelle SS427s. Yenko even converted 30 lightweight Novas with COPO-procured 427s, and later admitted that the result was "a

For 1970, the Chevelle had a mild facelift, but lovers of muscle cars could still opt for a "Turbo Jet 396" option, now actually measuring 402 cu inches (6.6 liters).

beast, almost lethal," capable of reaching 60mph in less than 4 seconds. The insurance companies agreed: they refused to cover the Nova 427, and no more were converted.

Chevrolet's Monte Carlo could not have been more different. Launched in 1970 as a luxury two-door, it was really an extended Chevelle. From the muscle point of view, the Monte Carlo's most significant feature was its SS454 package, which used Chevrolet's new 454-cu in (7.44-liter) version of the Turbo Jet big-block V8. This came in a mild state of tune – a 10.25:1 compression ratio and a single Rochester four-barrel carburettor – but still pushed out 360hp (268.5kW) at a relaxed 4,400rpm, plus 500lb ft (678Nm) at 3,200rpm. Still, the Monte Carlo was not really marketed as a muscle car, and this explain in part at least why only 2.6 per cent of them were ordered with the SS package and big motor.

In that same year, Chevrolet did the obvious thing and slotted the 454 into the Chevelle. It was obvious because the car had built up quite a reputation as one of the leading muscle cars, and even the top 396s were starting to get left behind by the 400-hp (298-kW) opposition. Chevelle SS454 buyers had a choice of two tunes: they could have the motor in its soft Monte Carlo form, nice and lazy, with bags of torque at low revs. But of more interest to the muscle car crowd was the LS6 version, with a higher 11.25:1 compression , bigger 780-cfm (22.09-m^3/minute) Holley four-barrel carburettor, big exhaust valves, and solid lifters. All told, it came to 450hp (335.5kW) at 5,600rpm and had the same peak torque as the softer LS5. To cope with the extra power, there were four-bolt main bearings, nodular iron bearing caps, and heavy

OPPOSITE and OVERLEAF: By 1972, the "396" tag had finally gone, replaced by the SS400.

duty con rods. Together, they could shift the Chevelle to 60mph (96.5km/h) in just 5.4 seconds, and turn in a quarter-mile in the high 13 seconds. It is little wonder that collectors see this as the ultimate Chevelle muscle car.

But this was the turning point. Insurance problems with high-powered muscle cars, in combination with new emissions and safety legislation, meant that the days of cars like the SS454 were numbered. For 1971, Chevrolet took the significant step of making the SS package (the stripes, suspension, and show-off bits) available on lower-powered 350 and 400 V8s as well as on the 454. It was the right move, as over three-quarters of SS buyers chose the smaller engines that year. By 1972 one could even get a Chevelle SS307, while the 450-hp 454 had gone. The remaining 454 was detuned, with lower compression and 270hp (201kW) net, but Chevrolet sold only 2,500 of this model in 1973. This was no longer what people wanted.

PLYMOUTH BARRACUDA & DODGE DART

Plymouth's Barracuda of the late 1960s and early 1970s, the hell-raising Hemi-powered 'Cuda that was tricky to insure, was a quintessential muscle car. But the original Barracuda was not like that at all. It started out in 1964 as a mild-mannered fastback version of the Valiant sedan. "We are sure," said Plymouth's general manager P.N. Buckminster, "that the Plymouth Barracuda is just right for young, sports-minded Americans who want to enjoy the fun of driving a car that also fulfills their general transportation needs." And that summed it up. The first Barracuda was a wholesome, sensible, fun car, though its degree of wholesomeness depended on what one got up to on the fold-down rear seat: "7 feet of fully carpeted 'anything' space."

Although it was Plymouth's response to the Mustang, the Barracuda was in no way a hot car. There was just one V8 option, a 273-cu in (4.47-liter) unit with single two-barrel carburettor and mild 180hp (134kW), though a four-speed manual gearbox with Hurst shifter was optional. This was not what mid-1960s

OPPOSITE & BELOW: 1970 Plymouth Barracuda. Not all Barracudas were Hemi-powered – most had milder, more insurable options such as this 383.

America wanted, and sure enough, only 23,000 Barracudas were sold in their first eight months.

But Plymouth realized what was holding back sales and rushed a revamped 'Cuda into production. Now the 273 V8 came with four-barrel

carburettor, a 10.5:1 compression ratio and 235hp (175kW). The four-speed box was still an option, but could now make the car sprint to 60mph (96.5km/h) in 8.2 or 9.1 seconds, depending on which magazine was doing the driving. To go with the hotter V8, a Formula S package brought wide wheels and tires, stiffer

In the early 1970s, showing off was definitely in, which would account for the bright orange livery, bold black stripes, and wide wheels of this Barracuda.

suspension, and rally stripes. This was more in tune with the market, and over 64,000 were sold.

That success encouraged Plymouth to make the Barracuda a model in its own right for 1967, and introduce a still-more powerful engine, this one a 383-cu in (6.28-liter) V8 pushing out 280hp (209kW) with its Carter four-barrel carburettor and a 10.0:1 compression. Front disc brakes, bucket seats and Sure-Grip differential were among the options, plus of course the inevitable rally stripes. Once again, over 60,000 Barracudas were sold, making these warmed up versions more popular than the hot and hairy 'Cudas ever were.

The Barracuda grew out of the Valiant, Plymouth's compact car. Fellow Chrysler division, Dodge, had its own compact, the Dart, and this became a formidable mini-muscle car for a few years in the late 1960s. Announced in 1968, the Dart GTS (it stood for GT Sport) was Dodge's equivalent of the Plymouth Road Runner, though unlike that mini-muscle, it never received the legendary Hemi power unit. But as it turned out, what it had was quite enough.

GTS buyers had the choice of two V8s. The 340-cu in (5.57-liter) small-block had a bore and stroke of 4.04 x 3.31 in (102.6 x 84.1mm); with 10.5:1 compression and a single four-barrel carburettor it could push out a respectable 275hp (205kW) at 5,000 rpm and 340lb ft (461Nm) at 3,200rpm. Alternatively one could have a 383-cu in (6.28-liter) big-block with 300hp (224kW), which was a lot of power for a compact. To go with the big V8, the GTS brought what Dodge described as Rallye suspension, 14 x 5.5-in wheels, and Red Streak tires; most cars were also ordered with a four-speed manual gearbox with Hurst shifter, or a competition-type TorqueFlite automatic. And being part of the Dodge "Scat Pack" (applied to the Charger R/T and Coronet R/T as well as the GTS) there were rear-end bumblebee stripes and hood power bulges.

There was little need for the little Dart to persuade anyone it was quick: 6 seconds was claimed for the 0–60-mph (0–96.5-km/h) sprint, and 15.2 seconds for the quarter-mile. If anything, those claims were slightly modest, as *Hot Rod* magazine recorded a 14.38-second quarter-mile in a TorqueFlite GTS. It got even faster in 1969, when the 383 V8 was uprated to 330hp (246kW), though the GTS remained a minority interest, with less than 7,000 (both hardtops and convertibles) sold in that year.

OPPOSITE: The 1968 Dodge Dart GTS was more of a budget muscle car.

BELOW & OVERLEAF: The 1970 Dodge Dart Swinger, a true budget muscle car, with high performance and low price.

More popular was the Dart Swinger 340. This was an unashamed budget muscle car that gave performance great priority and luxury very little. "Dart Swinger 340," went the advertisement, "Newest member of the Dodge Scat Pack. You get 340 cubes of high-winding,

4-barrel V8. A 4-speed Hurst shifter on the floor to keep things moving. All other credentials are in order." And they were: the Swinger had just as much power and performance as the GTS, not to mention the Rallye suspension, wide wheels, and bumblebee stripes. One could not have a convertible, and full carpeting only came if the four-speed was specified.

But who cared? At less than $3,000 the Swinger gave more performance per dollar than almost anything else, and Dodge sold 20,000 of them in 1969. Nor was a low price the only attraction: the Swinger proved cheaper to insure than other muscle cars with the same performance, as the insurance companies considered it a true compact car, and thus a better risk. Presumably they had not seen the performance figures: 6.5 seconds to 60mph (96.5km/h) and the standing quarter-mile in around 14.5 seconds could be expected. Scat Pack, indeed.

By 1971 the Dart had moved on. The year before, Plymouth had enjoyed great success with the Valiant Duster compact, and now it was Dodge's turn. So the 108-in (2.74-m) wheelbase coupé variant became a new Dodge for 1971, the Demon. Economy-minded motorists could order one of these with a 198-cu in (3.24-liter) six, but of more interest to muscle car fans was the small-block 340 V8, which took on that 275-hp motor that had done such good service in the Dart GTS and Swinger. The add-ons were much the same: standard three-speed manual transmission, with a four-speeder or TorqueFlite auto optional. Heavy-duty suspension, Rallye instrument cluster, E70-14 tyres, stripes and dual exhaust were all part of the package as well. If they did not make a big enough statement, then the optional black hood with two gaping air scoops, a rear spoiler or 'Tuff' steering wheel, surely would.

Motor Trend tested a Demon 340 against a Mercury Comet GT, Chevrolet Nova SS 350 and AMC Hornet SC360. The Demon was the heaviest of the lot, at 3,360lb (1524kg), but also the quickest, with a standing quarter-mile time of 14.49 seconds

and a time of 6.5 seconds to 60mph (96.5km/h). For 1972, the 340's compression
was lowered to 8.5:1, though the power drop to 240hp (179kW) was not as drastic
as it seemed: it was now measured on the SAE net rating. Whatever, it still made an
affordable muscle car, and was certainly easier to insure than its bigger brothers.
Interestingly, the car carried on into 1976, but the "Demon" name did not. Religious
groups took exception to it, and from 1973 the Demon was renamed the Dart Sport.

FORD: COUGAR & 427S

One would have expected Mercury, maker of luxury, upmarket Fords, to have been above the whole muscle car thing, but not a bit of it: there were performance Mercuries right through the 1960s. They may not have been as flashy as some (all those stripes, wings, and spoilers were not really in Mercury's marketplace), but they were there. Take the Comet Cyclone, a powered-up version of the standard Comet. For 1966 the Cyclone GT (based on Ford's Fairlane) was selected as the Indy 500 pace car (a sure sign of performance pretensions). Powered by Ford's 335-hp (250-kW) 390-cu in (6.4-liter) V8, complete with handling package and front disc brakes, it could do the quarter-mile in the high 14 seconds. So it was fast enough, though only around 16,000 were sold.

The Cougar, launched in 1967, promised to have more mass appeal. *Car Life* magazine described it as a "Mustang with class," and that is exactly what it was. The Cougar was based firmly on the pony car floorpan, but with its own bodyshell and tweaked suspension to provide a more comfortable ride. As far as buyers of muscle cars were concerned, the most serious Cougar was the GT 390. This made use of the same 390-cu in V8 as the Cyclone GT, and in the same relatively mild state of tune with three- or four-speed manual gearboxes or a three-speed Merc-O-Matic. Having softened up the suspension for standard Cougars, Mercury firmed it back up for the GT: stiffer springs, beefier shocks, and a larger front anti-roll bar, plus power front disc brakes, and wider tires. But the Cougar was not particularly

OPPOSITE & OVERLEAF: The Cougar was very much a performance car, an upmarket version of the Mustang, with the same performance options but a more luxurious interior.

fast (8.1 seconds to 60mph/96.5km/h and 16 seconds for the quarter-mile) and the fact that almost half were sold with the three-speed manual or auto suggested that many were being bought by non-sporting customers.

Things got more serious in 1968, with the 427-cu in (7.0-liter) GT-E, giving 390hp (291kW) in a fairly mild state of tune, rapidly superseded by the 428 Cobra Jet. Ford actually quoted only 335hp (250kW) for the new engine, but this was probably a ploy to make the performance Fords and Mercuries cheaper to insure. It is more likely that power was around the same as that of the 427 it replaced.

Mercury was now getting serious about racing, and entered the Trans Am series at around this time. One spin-off was the XR-7G : the G stood for Dan Gurney, and in fact "signature" special editions like this were popular at the time. The XR-7G was really a cosmetic package, and could be had with any of the Cougar power units.

By 1970, Mercury's slogan for the Cougar and its new Eliminator fastback derivative was "America's most completely equipped sports car." In keeping with the Mercury badge, it was marketed as a luxury sports car, which by and large it was. There was more noise insulation than in a Mustang, and a better-equipped

interior. Although the engine range was the same as that of the Mustang (including the high-revving Boss 302, the Cleveland 351, 428 CJ and even the 375-hp/279.5-kW Boss 429), the interior was pure Mercury, especially on the XR-7: vinyl high-back bucket seats with leather accents, map pockets on the seat backs, burr walnut effect on the instrument panel, loop yarn nylon carpeting, electric clock, tachometer, rocker switch display, rear seat armrest, map, and courtesy lights – so it went on.

For 1971, Cougar grew bigger and fatter (it was already larger than the original Mustang), and had now abandoned its pony car roots. Available as the base and sporty XR-7, choices of Cougar engine started with a two-barrel 351-cu in (5.75-liter) motor, offering 240hp (179kW). The same engine with a four-barrel carburettor produced 285hp (212.5kW), while the top-range 429-cu in (7.0-liter) motor claimed 370hp (276kW). It was a far cry from the Mustang, or indeed any other muscle car: it was heavier, with as much emphasis on trim and appointments as sheer performance.

That explained why the Cougar was never a major player in the muscle car market, and as the 1960s wore on the same was true for full-sized cars. They were too heavy and unwieldy to compete with a GTO or LS6 Chevelle, the whole basis of the muscle car concept being to squeeze a big engine into a relatively small car. Big engine in big car did not have the same effect.

Throughout the 1960s, however, the bigger Fords were able to turn in fast straight-line speeds as a result of one engine, the 427-cu in "side oiler." This had first appeared in 1963, primarily for NASCAR and drag racing, but Ford also built nearly 5,000 big Galaxies that year with 427 power. They came either in 425-hp (317-kW) street-tune form but with dual four-barrel Holley carburettors, or milder 410-hp (307-kW) form with a single carburettor. Whether for weekend racing or pure road use, in the base Galaxie or 500 XL, the new 427 gave a good account of itself. Heavy the Galaxies may have been, but from the way they scorched off the line many thought

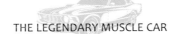

that 425hp was a conservative estimate. Ford itself claimed a 14.9-second standing quarter-mile for the 500XL hardtop 427 in 1965. By now the engines had names: Thunderbird High Performance and Thunderbird Super High Performance for the 410- and 425-hp units respectively.

Until 1966, the 427 could be had only in the full-size Galaxie, but for that year the mid-sized Fairlane was redesigned to make room in the engine bay for this big-block. Only about 60 Fairlane 427s were produced in 1966, but unofficial claims of a 14.5-second quarter-mile heightened the anticipation. As it happened, even in 1967 the 427 remained a limited-production option on the Fairlane, most GTs and GTAs leaving the line with Ford's familiar 390-cu in (6.4-litre) unit. Meanwhile, the new Galaxie "7 Liter" received Ford's new 428-cu in (7.0-liter) motor in 345- and 360-hp (257- and 268.5-kW) forms, though one could still pay extra and order the hot 427 in 410- or 425-hp (306- and 317-kW) forms.

From the power point of view, that was the high spot: for 1968, the 427 was detuned to 390hp (291kW), and production ended in the middle of that year. The newer Cobra Jet 428 and Thunder Jet 429 were certainly powerful, but could not compare with the original 427. Ford's mid-sized muscle-car flag was taken up by these engines in the Torino, the sporty version of the Fairlane 500. That came in stripped-down budget form for $2,699, and although on paper it produced a relatively paltry 335hp (250kW), Ford still claimed a 14.5-second quarter-mile as a result of the fact that the Torino was lighter than the big Galaxie. The fastback Talladega Torino, unveiled in the following year to bring NASCAR honours, used the same CJ-428 engine. For 1970, the new 429 motor replaced the 428, promising 360 or 370hp (268.5 or 276kW) depending on tune. But the vast majority of buyers went for the cheaper 302-powered car. It was surely a sign of the times.

CHEVROLET CORVETTE

Can the Corvette really be classed as a muscle car? Conventional wisdom includes V8-powered mid-sized sedans and compacts, or 2+2 pony cars, but not two-seater sports cars. Well, the Corvette was a V8 too, and it always concentrated on delivering a high performance per dollar rating, a key factor in the muscle car concept. And over a production life of 40 years it remained faithful to the front engine/rear drive layout, despite endless speculation about mid-engined prototypes. Straight-line performance via American V8 power was always part of that Corvette story: this, more than anything else, makes it a muscle car.

Like the four-seat muscle cars, the Corvette had its performance heyday in the 1960s, and each of its five generations featured a performance flagship. So a 327-hp (244-kW) fuel-injected small-block motor was offered as early as 1962. Then came the dramatic-looking Sting Ray of 1963, in open-top or twin-window convertible form. That car, with the ultimate L84 V8 with Ram Jet fuel injection, offered 375hp (279.5kW) at 6,200rpm, and 350lb ft (474.6Nm) of torque. Big-block Corvettes, the 396 and 427, soon followed, the latter in 390- and 425-hp (291- and 317-kW) forms, allowing 14-second quarter-miles and 0–60mph in less than 6 seconds. Most awesome of all was the special all-aluminium 427 fitted to just two Corvettes in 1969. This was the ZL-1, complete with ultra-high 12.5:1 compression, big valves, and 850-cfm (24.07-m³/minute) Holley four-barrel carburettor. For 1970, there was

OPPOSITE: The Corvette was a two-seat muscle car: a 1962 racing version is shown here.

OVERLEAF:The classic Corvette, the Sting Ray was available with fuel injection and over 400hp.

the biggest-yet 454-cu in (7.44-liter) V8 and high-compression, high-tune, high-revving 350 small-block at 390 and 370hp (291 and 276kW) respectively. Then came the 1970s and, like every other muscle car, the Corvette was forced to retrench and peg back its power to suit the less hedonistic times. But after a while, power outputs began to creep back up again, as performance came back into fashion and new technology allowed the combination of high power and lower emissions. Between 1987 and 1991 GM offered a 345-hp (257-kW) twin-turbo

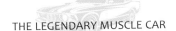
OPPOSITE: 1967 Corvette, this one showing the double-bend rear window.

BELOW: For 1968, the Sting Ray coupé gained a split rear window.

Corvette, the actual conversion completed by Callaway Engineering.

Chevrolet has continued to manufacture sports cars with the Corvette badge up to the present day.

INDEX

INDEX

PHOTOGRAPHIC
ACKNOWLEDGEMENTS
All images are supplied by Garry
Stuart. Other than the following:
©iStock.com and the following
photographers.
Page 3: Anton_Sokolov
Page 9: Thorsten Harries
Pages 10, 222: Brian Sullivan.
Pages 48, 124, 142, 143, 216, 235:
 Schlol.
Pages 79, 96, 102, 110, 154: Stan
 Roher.
Page 249: JSheets19